AF601558

Limnological Study of Tighra Reservoir

Dr. Dushyant Kumar Sharma

Professor of Zoology
Govt. Model Science College,
Gwalior, M.P.

2017

Scholars World

A Division of

Astral International Pvt. Ltd.

New Delhi – 110 002

© 2017 AUTHOR

Publisher's Note:

Every possible effort has been made to ensure that the information contained in this book is accurate at the time of going to press, and the publisher and author cannot accept responsibility for any errors or omissions, however caused. No responsibility for loss or damage occasioned to any person acting, or refraining from action, as a result of the material in this publication can be accepted by the editor, the publisher or the author. The Publisher is not associated with any product or vendor mentioned in the book. The contents of this work are intended to further general scientific research, understanding and discussion only. Readers should consult with a specialist where appropriate.

Every effort has been made to trace the owners of copyright material used in this book, if any. The author and the publisher will be grateful for any omission brought to their notice for acknowledgement in the future editions of the book.

All Rights reserved under International Copyright Conventions. No part of this publication may be reproduced, stored in a retrieval system, or transmitted in any form or by any means, electronic, mechanical, photocopying, recording or otherwise without the prior written consent of the publisher and the copyright owner.

Cataloging in Publication Data--DK
Courtesy: D.K. Agencies (P) Ltd. <docinfo@dkagencies.com>

Sharma, Dushyant Kumar *(Professor of zoology)*, **author**.
Limnological study of Tighra Reservoir / Dr. Dushyant Kumar Sharma.
pages cm
Includes bibliographical references and index.
ISBN 9789390384273 (Hardbound)

1. Limnology--India--Tighra Reservoir. I. Title.

QH183.S53 2016 DDC 551.48209543 23

Published by : **Scholars World**
A Division of
Astral International Pvt. Ltd.
– ISO 9001:2008 Certified Company –
4760-61/23, Ansari Road, Darya Ganj
New Delhi-110 002
Ph. 011-4354 9197, 2327 8134
E-mail: info@astralint.com
Website: www.astralint.com

Limnological Study of Tighra Reservoir

The Author

Dr. Dushyant Kumar Sharma is working as Professor of Zoology at Govt. Model Science College, Gwalior, M.P. He has been working in Department of Higher Education, Government of Madhya Pradesh since 1993. He did his B.Sc. Zoology (Hons) from Hindu College, University of Delhi, Delhi and received his M.Sc. and M. Phil degrees from University of Delhi, Delhi. Dr. Sharma obtained his Ph. D degree from Dr. B.R. Ambedkar University, Agra, U.P. Besides teaching, he is actively engaged in research. His main areas of interest are biochemistry, microbiology, immunology, hematology and environmental sciences. He has authored books on *Biochemistry, Microbiology* and *Immunology* and edited books on *Biological Diversity, Current Research in Environmental Sciences* and *Environmental Stress and its Remedies*. He is a co-author of two books on *Civil Services Preliminary Exams*. He has contributed several research papers in various national and international journals of repute. He is a member of Editorial Boards of *Indian Journal of Scientific Research* and *Indian Journal of Life Sciences*. Dr. Sharma is a life member of *The Indian Science Congress Association* and *Global Academic Society of India.*

Dedicated to
my Father
Late Shree L.P. Sharma

Preface

Each of one us knows the importance of water in life. It is a well known fact that *'water is life.'* The essence and sustenance of life depends on water. Water is a vital part of our life. We cannot think 'a life without water.' Besides drinking purpose, water is required for a variety of activities like household purposes, agriculture and industries.

We have been gifted with a large number of water bodies. But the careless attitude and uncontrolled race of men, in the name of development, have badly affected water and water bodies. Assessment of water quality is very essential for its proper utilization. A limnological study of Tighra Reservoir, Gwalior, Madhya Pradesh was carried out, under UGC sponsored Minor Research Project, by me. I am highly grateful to UGC (Central Regional Office), Bhopal for sanctioning me the project and providing me an opportunity to carry out the work.

The dream of publishing my work, in the form of a book, has been possible due to the keen interest and sincere efforts of Astral International (P) Ltd., New Delhi. I am thankful to my publisher and the whole production team for publishing this book in such a nice form.

During the course of my study and publication of the book I got help and guidance from many of my friends and colleagues. I express my sincere regards to them. Their suggestions have always helped me in improving my work. In addition, many people have assisted me in many other ways during the course of my study. I am thankful to each and every one who has been associated with me, directly or indirectly.

I wish to express my gratitude to my parents for their everlasting love and blessings. Finally, I acknowledge my thanks to my wife and children for their limitless personal support.

I hope that the book will be useful for the students, teachers and researchers and help them in their understanding of water and water bodies in a more fruitful manner.

Professor Dushyant Kumar Sharma

Contents

List of Figures

List of Tables

1
Introduction

Life and water are said to be the two faces of the same coin because all forms of life depend upon water for their existence. We cannot imagine life without water. It is rightly said that *'water is life.'* Man has been associated with water for a long time immemorial. Human civilization originated, developed and thrived in places in the vicinity of water. Water is an indispensible and valuable commodity. It is essential for the maintenance of life. It not only forms a major ingredient of living protoplasm but also helps several processes being carried on in the organisms. It is needed for human domestic consumption, agriculture, hydel power and industries.

About 70 per cent of the earth surface is covered with water. Over 99 per cent of this is in the form of oceans. Freshwater is one of the most precious yet undervalued resources. Although it may seems like a renewable resource, freshwater is a finite resources on earth.

In recent years, with the growing quest for industrialization and urbanization, there has been a great threat to our water resources. Freshwater is facing all round degradation all over the world. Deterioration of water quality, due to human activities, loss of aquatic biodiversity and depletion of water resources are the main challenges which need urgent attention. Dumping of all kinds of wastes - sewage, detergents, industrial wastes, pesticides and fertilizers, without thinking of their harmful impacts, have ruined the freshwater habitats. Domestic sewage creates a major problem. Organic wastes from domestic water cause problem of eutrophication, making water unsuitable for consumption. With increasing pollution, availability of pure water is decreasing at an alarming rate. Today, very rare or no river has been spared from industrial discharge. The agricultural practices, too, are greatly responsible for deterioration of water quality. Fertilizers, pesticides, animal wastes and sedimentation cause agricultural water pollution. This has greatly polluted water and rendered it useless for many purposes. Increasing pollution of water bodies has become a matter of great concern over a period of last few decades. In the present situation, it is necessary to have methods which should go parallel to the development, so that the water resources become environmentally sustainable. There is also a need to develop water resources in a way to get economic output in the price of aquaculture and fish culture. To achieve these goals, the first step is to study the water in all its aspects.

Freshwaters are biological systems and biogeochemical processes control the quality of freshwaters. Freshwater environment is highly diversified and marked by a wide range of physico-chemical conditions which greatly influence the life in water. The quality of any water resource is measured in the form of its physico-chemical parameters. The physico-chemical properties of water decide the quality of water and its biological diversity. The changes in the physico-chemical parameters tend to change the living conditions, especially in the number, diversity and distribution of the biota of that ecosystem. Fluctuations in physico-chemical factors adversely affect the organisms, limiting their production and interfering in the physiological processes which reduce their ability to compete with other populations within the environment.

The physico-chemical properties are not only affected by each other but they are also influenced by metrological conditions and biological parameters.

The physico-chemical analysis is the prime consideration to assess the quality of water for its best utilization like drinking, irrigation, fisheries and industrial purposes and is helpful in understating the complex processes and interactions between climatic and biological processes in the water.

For analyzing a water body, it is essential to study it in its all aspects- physical, chemical and biological. The effect of physical factors such as light and heat are of great significance as they are solely responsible for certain phenomenon such as thermal stratification, chemical stratification, diurnal and seasonal variations in the distribution and quality of planktons and other aquatic organisms. Planktons act both as producers and consumers and play important roles in the transformation of energy from one trophic level to another.

Thus, in order to make the maximum utilization of water bodies, it is essential to study the water in its whole aspect. The study is crucial for effective management and addressing practical problems of water quality and how water quality is influenced by human induced changes.

Limnology is the study of freshwater in all its aspects. The word 'limnology' is derived from Latin words *'limnos'* meaning lake and *'logos'* meaning study or discourse. Thus, limnology means study of lakes in the broadest sense. It deals with the study of all types of freshwaters including lakes, reservoirs, ponds, pools, rivers and streams. The study is also known by some other terms such as *freshwater biology, aquatic biology* and *aquatic ecology*. The study is much more important due to indispensability of water to mankind. According to Wetzel (2001) *'limnology is the study of the structural and functional interrelationships of the organisms of inland waters as they are affected by their dynamic physical, chemical and biotic environments.'* It encompasses an integration of physical, chemical and biological components of aquatic ecosystems.

Limnology is an interdisciplinary science, combing all aspects of hydrobiology, hydrochemistry, hydrophysics and geology. Limnological knowledge is essential not only for sound management of our aquatic resources but also more importantly for the development of aquaculture industry. It is emerging as one of the most important subjects since the lentic ecosystems hold more than 90 per cent freshwater used by man. Further, lake based primary and secondary productivities support aquaculture and fishery, contributing substantially to food security and fight against protein deficiency and malnutrition.

Earlier limnological studies were mostly descriptive and observational in nature. Many subsequent studies related to listing of species, animals, planktons, macrophytes, periphyton and nekton in aquatic habitat. But in recent years, studies have been ushered in the functional aspects within the aquatic ecosystems. The study of functional aspects involves the study of energy flow and nutrients cycling that control photosynthesis, food webs, food cycles and decomposition.

Many workers have made limnological studies of a large number of freshwater bodies throughout the world. Indian subcontinent with its great strategic position represents unique and varied physical, geographical and metrological features. The numerous freshwater habitats are represented by large rivers, streams, lakes and small ponds. They present a variety of features because of their distribution ranging from high altitude to plateau of our country. But unfortunately 70 per cent of these water bodies have gone polluted. There are a large number of seasonal, perennial ponds and lakes, located in rural areas in different parts of country.

In India, much work has been done on many aspects of limnology. However, very little work has been done on the impact of dynamic of climatic, physical and chemical parameters on planktons and productivity.

Keeping aforesaid facts in mind limnological studies of Tighra reservoir, the life line of Gwalior (M.P.), were under taken to investigate the physico-chemical and biological aspects. Tighra reservoir is the major source of drinking water to Gwalior city. Besides, the water of Tighra reservoir is also used for irrigation and pisciculture.The study was made to assess the hygienic and sanitary quality of the water and also to explore the flora and fauna of the reservoir. The study was carried out at four different sampling stations for one complete year. Planktons - both phytoplanktons and zooplanktons were studied qualitatively and quantitatively and interactions among biotic and abiotic factors were determined.

The study is an attempt to accumulate information pertaining to various aspects of hydrobiology of the Tighra reservoir. The observational and scientific data on the physico-chemical and biological study of the reservoir would enable to explore new possibilities for better augmentation of aquaculture.

2
Review of Literature

A number of workers from India and different parts of the world have made great contribution in the field of limnology for a long time. Some of the legends in the field of limnology are Hutchinson (1941, 1957), Smith (1950), Welch (1952), Ruttner (1963), Hynes (1970), Wetzel (1975), Golterman (1975) and Pennak (1978).

Adebisi (1981) made limnological investigations of tropical river upper urgan river, Nigeria. Franco David (1987) studied limnological characteristics of Sanger Lake, USA. Aznar *et al.* (1991) investigated several physico-chemical parameters as well as saprophytic and public health related heterotrophic, bacterial groups in Albufera Lake, in Valencia Spain. Bhodra *et al.*(2003) observed marked seasonal variation in various physico-chemical parameters in the river Torsa of North Bengal. Kargul *et al.* (2005) statistically evaluated the seasonal changes of some water quality parameters of the Buyuk Melen river basin in Turkey. Medina –Junior and Rietzler (2005) carried out limnological study involving physical, chemical and biological aspects with emphasis on the zooplankton community of Pantanal saline lake in Brazil. Kamal *et al.* (2007) investigated total 22 the physico-chemical properties of Mouri river Khulna, Bangladesh. Correlation and the '*t*' values among the parameters were also determined by them. Elmaci *et al.* (2008) evaluated the physico-chemical and microbiological properties of lake Uluabat, Turkey, known for its scenic beauty and richness of aquatic life. Gurung *et al.* (2009) reviewed the limnological status of high altitude lakes in Nepal.

In India, there are several hundreds of inland water bodies such as ponds, lakes and reservoirs and rivers, located both in tropical and temperate regions.

Among early Indian workers, works of Ganapati (1940, 1949, 1950, 1956, 1960), George (1961, 1962,), Govind (1963), Zafar(1964, 1966, 1967) Sreenivasan (1964, 1965, 1968, 1976), Sahai and Sinha (1969), Munawar (1970), Vyas *et al.* (1977), Mishra and Yadav (1978) and Qadri and Yousuf (1980) are of worth mentioning.

Ganapati (1940, 1949, 1950, 1956 and 1960) studied the limnology of various reservoirs and tanks of south India. Zafar (1964, 1966) studied the limnology of Hussain Sagar Lake. Physico-chemical studies of Kamla Nehru Tank in Muzzaffar Nagar were carried out by Verma and Shukla (1970). Munawar (1970) investigated the hydrobiology of freshwater ponds of Hyderabad.

In the northern India, Gadded *et al.* (1983) made ecological studies of water bodies of Gulberga. Shukla *et al.* (1989) studied physico-chemical and biological characteristics of river Ganga from Mirzapur to Ballia and Chopra and Patric (1994) observed the effect of domestic sewage on river Ganga. In 1994 Singh and Singh made correlation between dissolved oxygen and effluent quantity in Ganga water. Joshi (1996) investigated the hydrobiological profile of river Sutlej. Joshi and Tyagi (1997) studied the limnological features of the Chirapani stream in Kumaon Himalaya. Mani and Gaikwad (1998) carried out physico-chemical analysis of Lake Pokhran.

Dutta *et al.* (2001) made hydrobiological studies of river Basantan Samba, Jammu. Jha and Barat (2003) studied the hydrobiology of Lake Mirik in Darjeeling, Himalayas. Bhutani and Khanna (2007) worked on the limnological status of Suswa river in Uttarakhand. Ram and Singh (2007) studied water quality of Ganga river with special reference to physico-chemical and bacteriological parameters. Sharma *et al.* (2008) assessed the water quality of Udaipur lakes by observing the physico-chemical and microbiological parameters. Bhat *et al.* (2009) recorded some physico-chemical parameters of some urban ponds of Lucknow, Uttar Pradesh. Shastri and Pendse (2001) investigated the hydrobiological study of Dahikhuta reservoir. Parray *et al.*(2010) investigated limnological profile of a sub urban wetland – Chatlam, Kashmir. Pokale *et al.* (2010) analyzed the water quality of Pench reservoir.

Many workers have explored the freshwater bodies of southern India and have made great contributions in the field of limnology. Agarkar *et al.* (1998) observed physico-chemical parameters of Kalbadevi estuary in Ratnagiri District of Maharasthra. Physico-chemical characteristics and pollution level of Nainital Lake were analyzed by Ali *et al.* (1999). They showed that the lake water was rich in nutrients which supported growth of many macrophytes and algal blooms. They also revealed the role of macrtophytes and phytoplanktons in bio-monitoring and phyto-remediation of toxic metal ions.

Yalavarthi (2002) made hydrobiological studies of the red hills reservoir, North Chennai, Tamil Nadu. Kamble and Angadi *et al.* (2005) studied the limnology of Papnash pond, Bidar in Karnataka. On the basis of their studies they revealed that the pond water was hard, alkaline and polluted. Mali *et al.* (2006) made physico-chemical analysis of Bhategaon reservoir in Parbhani district, Maharahtra. Joshi (2006) studied the physico-chemical analysis of water from Ekruk reservoir. Maniappa and Naik (2007) investigated the physico-chemical properties of Malaprabha river in Karnataka. They studied 25 different parameters to assess the quality of water. Seasonal variations in water quality of Priyar river basins were conducted by Muhamed Ashraf and Mukundan (2007) with a view to utilize the water for drinking purpose. Deswal and Chandana (2007) carried out physico-chemical and bacteriological analysis of water of a sacred Sannihit Sarovor in Kuruksshetra. Eswaralakshmi *et al.* (2008) investigated the physico-chemical factors of water of Vattambakkam Lake in Kanhipuram, Tamil Nadu. Kamble and Muley (2008) studied physico-chemical parameters of Kalbaddevi estuary. Umamaheswari and Anbusaravanon (2009) assessed the water quality of Cauvery river basin in Trichirappalli. Bade *et al.*(2009) studied physico–chemical parameters in Sai reservoir, Latur district, Maharashtra. Solaraj*et al.* (2010) studied the water quality of Cauvery Delta river basin. They found that monsoonal rains and subsequent increase in river flow rate influence certain parameters like dissolved solids, phosphates and dissolved oxygen.

Recently, significant work has been done by many researchers. Jagtap *et al.* (2011) observed seasonal variations in Sino-Kolegaon reservoir in Osnanabad, Maharashtra. Ahirrao and Pedge (2011) analyzed the water quality of Tridhara, a holy place in Parbhani district, Maharashtra. In their studies they recommended that the water should not be used for drinking or bathing purpose. Sawant *et al.* (2011) analyzed the physico-chemical properties of the Uttur Tank in Ajara tahsil of Maharashtra.

Thakur *et al.* (2011) studied the seasonal variations in different physico-chemical parameters of Pariyej Lake and found them within the permissible limits of WHO. Pandey *et al.* (2011) investigated the water quality of river Tons at Akbarpur, U.P. Sinha and Biswas (2011) analyzed the physico-chemical characteristics of a lake in Kalyani, West Bengal.

Attention on lakes and reservoir from Madhya Pradesh enlightens the work of past three decades. Many workers have observed physico-chemical and biological factors of several water bodies of M.P.

Mention may be made of Adholia (1979, 1992), Sharma and Adoni (1982), Nayak *et al.* (1982), (Unni, 1984, 1985), Saksena and Kulkarni (1986), Saksena and Mishra (1988), Palhaya and Malvia (1988), Mathew (1992), Unni *et al.* (1992), Mishra (1993, 2005), Prakash (1996), Poi and Shukla (1997), Sharma and Jain (2000), Sharma (2005, 2006, 2007) and Saksena *et al.* (2008).

Nayak *et al.* (1982) investigated seasonal variations in relation to physico-chemical factors of Matyatal, Panna.

Gwalior, located in the northern region of Madhya Pradesh, has several water bodies. Kaushik *et al.* (1991a) studied the water quality and phytoplanktonic algae in Chambal Tal. Dagaonkar and Saksena (1992) made physico-chemical and biological characterization of a temple tank in Gwalior.

Adholia (1992) worked on the bioenergetics of Mansarover reservoir in Bhopal with reference to fish production. Kartha and Rao (1992) made environmental studies of Gandhisagar reservoir. They observed marked seasonal variations in temperature, transparency, pH, DO, chloride, nitrate, phosphate and silicate.

Jain and Sharma (2000) studied the water quality of Rampur reservoir of Guna district to explore the possibilities for better management of pisciculture. Mahajan *et al.* (2002) studied Kunda River in Khargaon to investigate its physico-chemical and biological properties. Jain *et al.* (2002) worked on the ecology and fish fauna of Gopalpura Tank of Guna. Patel *et al.*(2002) made hydrobiological observation on Virla reservoir in Khargaon. In year 2006, Sharma observed seasonal variations in physico-chemical characteristics of Rampur reservoir of Guna. His study revealed that the water of the reservoir was quite suitable for pisciculture. Sharma (2007) studied the ecology and fish fauna of Makroda reservoir of Guna and recorded ten species of fishes from the reservoir.

Garg *et al.* (2010) studied the seasonal variations in water quality and major threats to Ramsagar reservoir, Datia.

Buktar and Sakhare (2011) analyzed the physico-chemical properties of Wan reservoir and found the water to be suitable for drinking purpose. Jagtap *et al.* (2011) carried out seasonal studies of Sina-Kolagaon reservoir in Maharasthra. Khaire *et al.* (2011) made an assessment of physico-chemical parameters of Mehakari reservoir and on the basis of their studies revealed that the water body was more productive. Seasonal variations in physico-chemical and microbiological properties of water of river Achencovil, Kerala were studied by Sanal Kumar *et al.* (2011). Banerjee and Ghosh (2012) conducted a study to assess the limnological characteristics of the river Damodar, West Bangal. Subramanian *et al.* (2012) investigated groundwater quality in Andimadam area, Perambalur district, Tamil Nadu. Water quality of Gomati river water in and around Sultanpur of Uttar Pradesh was studied by Srivastava and Srivastava (2013).

Recently, Dhar and Slathia(2014) studied seasonal water quality of Lake Mansar, Jammu. Shinde and Kolhe (2014) made comparative evaluation of some physico-chemical and microbiological parameters in Godavari river Basin at Tapovan area of Nashik. An assessment of lake water quality index of Manipu lake of district Ahmedabad, Gujarat was carried out by Kotadiya and Acharya (2014).

Biological Studies

Planktonic Studies

Planktons are the weakly swimming but mostly drifting small organisms that inhabit different water bodies. The planktons are divided into two groups such as phytoplanktons and zooplanktons.

Phytoplanktons include algae that live near surface where there is sufficient light to support photosynthesis. They are the autotrophs and constitute the primary producers of freshwater ecosystems. These organisms are the important biotic component of an aquatic ecosystem and provide the main food of fishes directly or indirectly and can be used as indicator of the trophic phase of the water body. Major phytoplankton groups present in the water are - Bacillariophyceae, Chlorophyceae, Myxophyceae and Euglenophyceae.

Zooplanktons are minute aquatic organisms that are nonmotile or are swimmers and they drift in water columns of a ocean, sea or freshwater bodies to move any great distance. They are heterotrophic in nature and play important role in food web by linking primary producers and higher trophic levels. The freshwater zooplanktons comprise of protozoa, rotifers, cladocerans, copepods and ostracodes.

The study of planktons of a water body is very important for proper manipulation of the factors influencing the biological productivity of the wetland. The ecology of planktons is helpful in opening a new vista in the field of aquatic ecology.

Many workers have made significant contribution in the study of biological parameters of freshwater bodies, both in India and abroad. Early work on planktonic studies includes that of, Ganapati (1940,1949), Nyygaard (1949), Smith (1950), Chacko and Krishnamurthy(1954), Khan and Siddiqui (1954), Chakarabarty *et al.* (1957), Green (1960), George (1961,1962), Govind (1963), Sreenivasan (1964,1966, 1976), Laxminarayana. (1965), Arora (1966), Zafar (1964, 1967), Michael (1968), Unni (1971), Nasar (1977), Jana and Das (1978), Singh and Sahai (1978), Rai (1978), Adholia (1979), Joshi *et al.* (1979), Jana (1979), Dutta *et al.* (1982,1983),Kundu and Sarkar (1987), Shaha and Pandit (1990),Kaushik *et al.* (1992), Pandey *et al.*(1992), Bais *et al.*(1993), Bilgrami *et al.*(1994), Singh *et al.* (1996), to name a few.

Davis (1955) studied the planktons and plankton production. Andersan *et al.* (1955) studied the relationship between phytoplanktons and zooplanktons in two lakes of Washington.

In Southern India, Chacko and Krishnamurthy (1954) studied the planktons of freshwater fish ponds in Madras city. Chakarabarty *et al.* (1957) made a qualitative analysis of the planktons of the river Jamuna. Sahai and Sinha (1960) investigated the bioecology of inland water of Gorkhapur.

Sharma and Bhatnagar (1977) recorded seasonal variations in planktons in Bhopal Lake. Prakash *et al.* (1978) recorded 5 species among protozoa, 12 species among rotifera and 7 planktonic species of oligochaeta, in their studies on river Jamuna. Adholia (1979) studied the zooplanktons of river Betwa in Madhya Pradesh. Kohli (1981) identified the planktons of Govind Sagar reservoir.The composition of the rotifer fauna in Morar and Swarnrekha rivers at Gwalior have been described by Saksena and Kulkarni (1986). Verma and Munshi (1987) studied the planktons of Budwa reservoir of Bhagalpur.

The density and diversity of zooplanktons are influenced by several physico-chemical factors. Temperature, dissolved oxygen and organic matter are the important factors which control the growth of zooplanktons as reported by Bharti and Rana (1987).

Studies on zooplanktons in relation to water quality of river Kshipra, in Madhya Pradesh, have been carried out by Kulshrestha *et al.* (1989). Kaushik *et al.* (1991b,1991c) studied phytoplanktons of water bodies of Gwalior region.

Adholia and Vyas (1992) determined correlation between copepods and limnochemistry of Mansarovar reservoir. Kumar (1995) studied periodicity and abundance of planktons, in relation to physico-chemical characteristics. Goel and Autade (1995) recorded 61 species of phytoplanktons and 19 zooplankton species in Panchganga River at Kolhapur. Joshi *et al.*(1996) studied planktonic population of Ganga canal, in relation to certain physico-chemical factors.

Khanna and Singh (2002) observed seasonal fluctuations in the planktons of Suswa River at Raiwala, Dehradun. Akin-oriola (2003) monitored zooplankton abundance, composition and environmental parameters in Ogunpa and Ona rivers, Nigeria.

Maruthanaryagam *et al.* (2003) studied season specific zooplankton diversity in Thirukkulam pond, Mayiladuthurai, Tamil Nadu. Sukand and Patil (2004) recorded four major groups of zooplanktons in their studies on Fort lake Belgaum, Karnatka. Rotifers constituted (52.38 per cent), followed by copepods 26.5 per cent, cladocerans 16.45 per cent and ostracodes 4.67 per cent. Jeelani *et al.*(2005) studied distribution and ecology of phytoplanktons in Dal lake. Adesalu and Nwankwo (2005) identified three classes of phytoplanktons – Bacillariophyceae, Cyanophyceae and Xanthophyceae in Olero Creek and Benin river, Nigeria. Chavan *et al.* (2005) studied seasonal variations of abiotic factors of Manjara Project water Reservoir in Beed, Maharashtra.

Kaur *et al.* (2005) studied the correlation between rotifers and physico-chemical factors at Harika reservoir in Punjab. Aher and Nandan (2005) made an assessment of water quality of Mosam river of Maharashtra with relation to phytoplanktons. Bhat and Pandit (2005) studied the phytoplanktons of Anchar River, Kashmir. They recorded 143 species of algae. Jeelani *et al.* (2005) recorded a total of 127 phytoplanktons from Dal Lake, Kashmir. Phytoplankton population in Dal lake consisted of six major groups namely Bacillariophyceae, Chlorophyceae, Cyanophyceae, Dinophyceae, Euglenophyceae and Chrysophyceae. Tiwari and Chauhan (2006) studied seasonal diversity of planktons of Kitham Lake in Agra.

Murugesan *et al.* (2007) observed the physico-chemical and biological properties of river Chittar at Courtallam, Tamil Nadu.

Shayetehfar *et al.* (2008) studied the effect of some physico-chemical parameters on surface water density of rotifers from Kor river, Fars, Iran. Senthikumar and Sivakumar (2008) studied physico-chemical parameters of Veeranam lake in Cuddalore district of Tamil Nadu in relation to phytoplankton diversity. Sreelatha and Rajalakshmi (2008) observed the zooplankton diversity of Goutami Godavari Estuary, Yanam, Pandicherry.

Nogueina Marcos *et al.*(2008) investigated zooplankton assemblage in a large tropical river in Brazil. Shekhar *et al.* (2008) studied the phytoplanktons to assess the water quality status of river Bhadra, Mysor with reference to industrial pollution. Hulyal and Kaliwal (2008, 2009) investigated the physico-chemical factors of Almatti reservoir of Bijapur district of Karnataka, in relation to zooplanktons and phytoplanktons.

Chattopadhyay and Barik (2009) worked on the zooplanktons of a freshwater lake at Burdwan. Both phytoplanktons and zooplanktons act as bio-indicators. Ferdous and Muktadir (2009) reviewed the potentiality of zooplanktons as bio-indicators.

Achionye-Nzeh and Isimaikaiye (2010) recorded the fauna and flora of a reservoir in Nigeria. Basavarajappa *et al.* (2010) discussed the phytoplankton diversity in Hadhinaru lake, Mysore.

Ingole *et al.* (2011) studied the water quality of Majalgaon Dam with special reference to zooplanktons. Soruba and Sangeetha (2011) identified the phytoplanktons of temple pond at Tiruvalluvar.

Vidya Gurkar and Mahesh (2011) studied the phytoplankton diversity in Varuna Lake of Mysore. Harish Kumara and Srikantaswamy (2011) worked on the seasonal variations of planktons of Tungabhadra reservoir. Singh and Laura (2012) made an assessment of physico-chemical properties and phytoplankton density of Tilyar Lake, Rohtak (Haryana). Panigrahi and Patra (2013) studied seasonal variations in phytoplankton diversity of river Mahanadi, Cuttack city, Odisha.

Recently, Quadri (2015) has studied the diversity of planktons in Kham river, Aurangabad. He investigated 20 species of zooplankton belonging to different groups *i.e.* Rotifera, Copepoda, Cladocera, Ostrocoda.

Productivity

Sreenivassan (1964, 1965 and 1976) studied the hydrobiology of some reservoirs and tanks of Madras city with special reference to primary production and fish production. Hussainey (1967) studied the primary productivity and plankton periodicity of a tropical lake. Jana *et al.* (1978) worked on the ecology and plankton production in fish ponds in West Bangal. Gopal *et al.* (1978) studied the primary production of different types of ponds at Bhagalpur. Sreenayya and Zafar (1979) reported the phytoplankton production in Mir Alam Lake of Hyderabad.

Palaniappan *et al.* (1981) studied two rock pools of Salum and revealed that the primary productivity of rock pools was dependent on the amount, duration and frequency of rainfall and also on the location of water bodies. Primary productivity of Ramasamudra at Manglore was studied by Ayyappan and Gupta (1985) who reported that the primary production was controlled by several hydro-chemical factors.

Khan *et al.*(1988) determined primary production in Sheikh jheel at Aligarh (U.P.).Velecha and Bhatnagar (1989) studied the production of phytoplanktons in eutrophic lake of Bhopal. Saha and Pandit (1990) worked on dynamics of primary productivity between lentic and lotic systems in relation to biotic factors.

Vijaykumar (1995) studied the primary productivity of Gulberga pond, Karnataka.

Kumar (1996a) investigated the impact of coal mining effluents on the primary productivity of water bodies in and around coal fields. In the same year, Kumar (1996b) also observed impact of sewage pollution on the chemistry and productivity of two freshwater bodies in Bihar. Bais *et al.* (1997) reported nitrite and phosphate, as the main factors, for the control of phytoplankton production in Sagar and Military engineering lake, Sagar, Madhya Pradesh, respectively.

Study of the primary production, in relation to physico-chemical nature of the water, was investigated by Singh (1998). Singh and Singh (1999) conducted a comparative study on the phytoplankton primary production of Ganga river and a pond at Patna, Bihar, India. Sarkar and Prakasham and Joseph (2000) worked on the water quality of Sasthamcotta lake in Kerala, in relation to primary productivity and pollution from anthropogenic source. A short-term study on certain physico-chemical factors and primary productivity of Viraganur dam, Madurai was conducted by Jeyachandran and Krishnan (2001). Das (2002) worked on phytoplankton primary production in some reservoirs of Andhra Pradesh. Sukumaran (2002, 2003) estimated the primary production of Lalbagh tank in Bangalore.

Raj Kumar (2004) analyzed the phytoplantonic primary production of a polluted freshwater pond of Krishnan Anaikatikulum in Pollachi, Coimbatore. Similarly, Mali and Gajaria (2004) studied the primary productivity of a fish culture pond in Gujrat. Kumar and Bhore (2005) studied the dynamics of phytoplankton productivity.

Role of limnological factors on phytoplankton productivity was investigated by Shiddamallayya and Pratima (2007) who concluded that there are several environmental factors such as local hydrography, seasonal variability and limnological factors which act simultaneously on the production capacity of the water body.

Shiddamallayya and Pratima (2007) investigated the role of limnological factors on phytoplankton productivity. Siva Kumar and Karuppasamy (2008) studied various factors which affect productivity of phytoplanktons.

Bhossle *et al.* (2010) observed seasonal variations in primary productivity of some selected lakes in Maharashtra.

Vasanthkumar and Vijay Kumar (2011) studied diurnal variation of primary productivity in Bheema River. They concluded that the rate of photosynthesis was greater in early hours of the day light and it decreased markedly in the afternoon.

The primary productivity of Bay of Bengal at Digha has been assessed both spatially and seasonally by Moharana and Patra (2013).

3
About the Reservoir

Gwalior, an ancient city known for the great musician Tansen, is situated in the north region of Madhya Pradesh. The city is gifted with a number of historical and tourist places. The Tighra reservoir, the life line of Gwalior, was primarily constructed to fulfill the water supply of the city. The reservoir is also used to culture fish by the fisheries department and for irrigation purpose.

Location

The Tighra reservoir is situated about 20 km west of Gwalior city, near Tighra village which is in close proximity of SADA Magnet city. It lies on 26°13′ N latitude and 78°30′ E longitude, at an altitude of 218. 58 m. The reservoir is surrounded by hills from three sides. The hills on the north and western side are 300 m high and those on southern and south east side are about 225m high.

At the south western side, river Sank joins the reservoir through a gorge. About a dozen of small nallahas drain in the reservoir from the hill slopes. In the north east of the reservoir there is a concrete masonry wall.

History

The construction of the reservoir was started in the year 1910 across a seasonal rain fed Sank river by Late Maharaja Madhav Rao Scindia and the work was completed in the year 1910.The reservoir was, then, known as Madhav Sagar. Later, the reservoir is being known as Tighra reservoir, after the name of the village Tighra, near which it is situated. The reservoir was reconstructed in the year 1929. From time to time, local administration and volunteer organizations take initiative for deepening of the reservoir so that the water level of the reservoir could be maintained for a long time.

Morphometry

The reservoir is irregular in shape having shallow embayment at its periphery (Figure 1). The maximum length of the reservoir along the south west - north east axis is 5.8 km and maximum width is 3.8 km. The maximum depth of the reservoir is 18 m. The reservoir has a live storage capacity of 124.23 m (4390 mcft), leaving a dead storage capacity below lowest still level of 6.56 m (232 m cft). The catchment of the reservoir is 160 sq. miles (414.24

Figure 1: Satellite Image of Tighra Reservoir (*Source* : Google earth)

Figure 2: A View of Tighra Reservoir.

sq. km). The water spread area of the reservoir is 7.5 sq. miles. Some salient features of the reservoir are summarized in Tables 1 and 2.

Table 1: Some Salient Features of Tighra Reservoir, Gwalior

Location	On Sank River
Nearest village	Tighra
Latitude	26° 13' N
Longitude	78° 30' E
Altitude	218.58 m
Catchment area	414.25 sq. km
Maximum length	5.8 km
Maximum width	3.8 km
Maximum depth	18 m
Capacity	
Gross	130.80 m^3 (4622 mcft)
Live	124.23 m^3 (4390 mcft)
Dead	6.56 m^3 (232 mcft)

Contd...

Table 1–*Contd...*

Principal levels	
LBL	202.5 m
Lower still	211.8 m
Crest	202.2 m
Full tank (FTL)	225.5 m

Table 2: Some Salient Features of the Dam of Tighra Reservoir, Gwalior

Type	Masonry dam
Length	1524.2 m (Masonary 1342.4 m and Earthen 182.8 m)
Maximum height	24 m
Estimated cost	52.27 lakhs
No. of overflow sluice gates	64
Area of gates	10' x 8'
No. of fitting gates	16
Area	10' x 4'
Discharge capacity of gates	
Tilling gates	260 cusecs each
Vertical gates	734 cusecs each
Irrigation canals	
Jisoli minor	6.4 km
Nirouli minor	4.8 km
Water supply to city	66.37 mld

Climate

The Tighra reservoir has very severe cold and very hot summer. The climate on the whole is mild hot. The rainfall is confined to late June to September end. The drainage of the reservoir is fed by the monsoon. The annual rainfall for the study period was 1103.9 mm.

Sampling Stations

For physico-chemical and biological studies of the Tighra reservoir, the reservoir was divided into four zones. In each zone one sampling station was

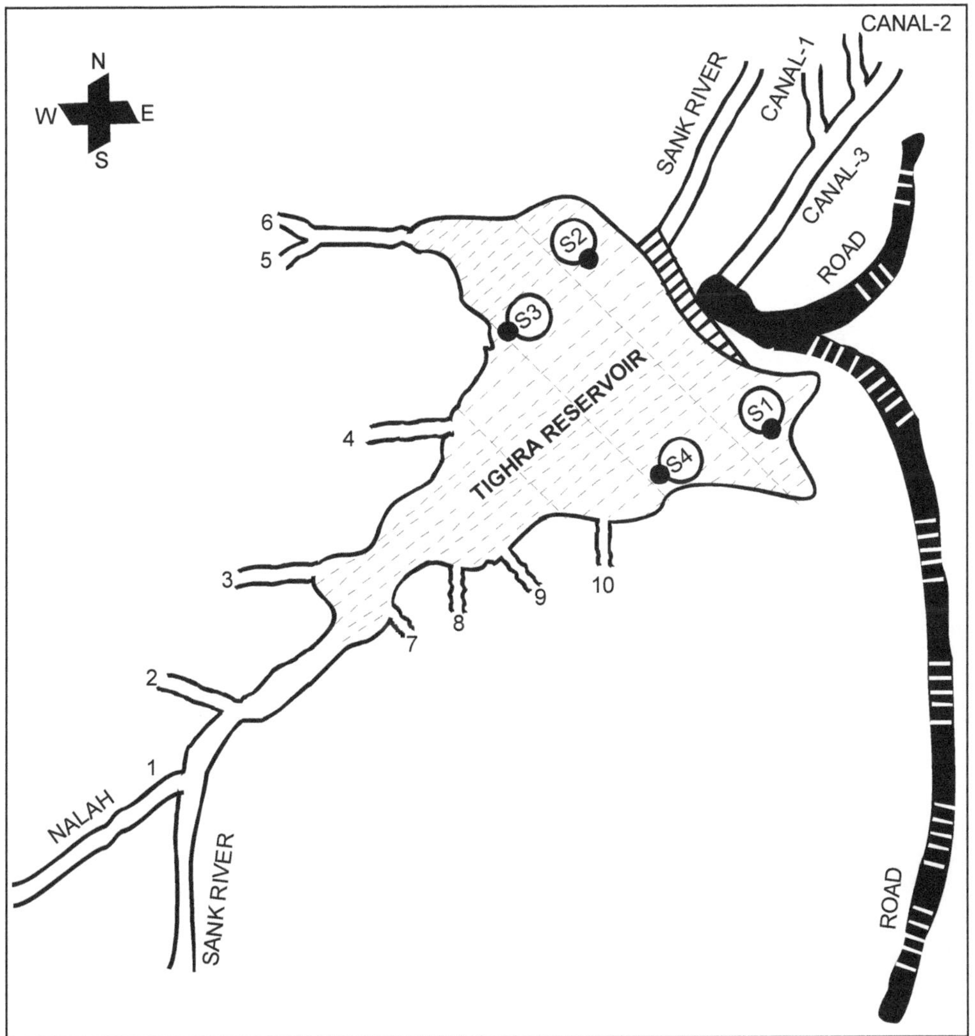

Figure 3: Different Sampling Stations Marked as as S1, S2, S3 and S4.

Sampling Station 1

Sampling Station 2

Sampling Station 3

Sampling Station 4

Figure 4: Different Sampling Stations.

selected, as marked in the Figure 3 as S1, S2, S3 and S4. The sampling stations were so selected as to cover the maximum area of the reservoir (Figure 4).

4

Materials and Methods

Physico-chemical Parameters

The water samples were collected on monthly basis from four different sampling stations - S1, S2, S3 and S4, during November 2010 to October 2011 in the morning hours. The samples were collected in 5 liter clean plastic cans. The pH and temperature of water samples were estimated on the spot, while other parameters were estimated in the laboratory, employing the methods described by APHA (1989), Trivedy and Goel (1986), Saxena (1998), and Khanna and Bhutiani (2008).

Physical Parameters

Colour

For determination of the colour, samples were collected in clean glass bottles and analysis was done soon after collection of the samples. Colour was determined by Visual comparison method.

50 ml sample was taken in Nessler tube. The colour of the sample was noted looking vertically through the tubes towards a white surface, placed at such an angle that light is reflected upward through the column of the liquid.

Standard colour solution was prepared by dissolving 1.246 g of potassium chloroplatinate and 1.0 g crystalline cobaltous chloride in distilled water having 100 ml. of conc. HCl to prepare 1 lit of solution. This solution had 500 mg/lit of platinum and about 250 mg of metallic cobalt and thus equivalent to 500 colour units.

Standard solution of 5, 10, 20, 25, 30, 40, 45, 50, 60 and 70 colour units were prepared by diluting the stock solution of 500 colour units with distilled water.

If the colour of the sample exceeded 70 units, sample were diluted with distilled water and colour units were calculated by the following formula:

$$\text{Colour units} = \text{Estimated colour} \times \text{Dilution factor}$$

Transparency

Transparency was determined by using Secchi disc method. For this purpose, Secchi disc was lowered in the water with the help of string, tied to it, until it just disappears. The depth was noted by marking on the string. Then Secchi disc was uplifted and the depth at which it reappears again was noted.

Transparency was calculated by the following formula:

$$\text{Transparency} = A+B/2$$

where,

A = Depth at which Secchi disc disappears.

B = Depth at which Secchi disc reappears.

Temperature

For determination of water temperature, water was collected and temperature was measured by a mercury thermometer and reading was noted down.

Turbidity

Turbidity is caused by the presence of debris and after rapidly settable matter. When light is passed through a sample having suspended turbidity, some of the light is scattered by the particles. The scattering of the light is generally proportional to the turbidity. The turbidity was measured by using Systronics Digital Nephelometer (Model : Type 132).

Conductivity

Conductivity is the measure of the ability of a substance to conduct electric current. It is due to the presence of various ions in the water. The conductivity is measured with the help of conductivity meter. The unit of conductivity measurement is siemens cm^{-1} or (μScm^{-1}). The conductivity was measured with the help of Labtromics Conductivity meter (Model No. LT-16).

Chemical Parameters

Free CO_2

Free CO_2 was determined by titrating the water sample using a strong alkaline NaOH. 100 ml of water sample was taken in a conical flask and a few drops of phenolphthalein indicator were added. If the colour turned pink, free CO_2 was absent. If the sample remained colourless, it was titrated against 0.05 N NaOH till a pink colour appears as the end point. Three readings were taken and average of them was used for calculating free CO_2 by using following formula:

$$\text{Free } CO_2 \text{ (mg/lit.)} = \frac{\text{ml} \times \text{ml N of NaOH} \times 1000 \times 4}{\text{ml of sample}}$$

Dissolved Oxygen

Dissolved oxygen (DO) was determined by using azide modified Winkler's method. The samples were collected in BOD bottles (100 ml) avoiding any kind of bubbling (trapping of air bubbles) in the bottles after placing stopper. 1 ml each of $MnSO_4$ and alkaline iodide azide solutions were added well below the surface from the walls. A precipitate appeared. The bottles were shacked properly and left to settle down for some time. 1-2 ml of concentrated H_2SO_4 was added and samples were shacked well to dissolve the precipitate. 100 ml sample was taken in a conical flask and titrated against 0.025 N sodium thiosulphate solution, using starch as indicator. At the end point, initial dark blue colour changes to colourless.

Three readings were taken and average of them was used to calculate DO using the following formula:

$$\text{DO (mg/lit)} = \frac{\text{ml} \times \text{N of sodium thiosulphate} \times 8 \times 1000}{V_1 - V}$$

where,

V_1 = Volume of sample

V = Volume of $MnSO_4$ and alkaline iodide azide

Total Hardness

Hardness, in water, is caused mainly by the calcium and magnesium ions. Total hardness was calculated by using EDTA method. Calcium and magnesium form a complex of wine red colour with Erichrome Black T at pH 10.00±0.1. The EDTA has a strong affinity towards Ca^{2+} and Mg^{2+} and, therefore, by addition of EDTA a wine coloured complex is broken down and a new blue colour is formed.

For determining total hardness 50 ml water sample was taken in a conical flask and 1 ml buffer was added. 100 to 200 mg Erichrome Black T indicator was added which turned the solution wine red. The solution was titrated against 0.01 M EDTA solution till the wine red colour changes to blue (end point). Total hardness was calculated by using the following formula:

$$\text{Total Hardness (mg/lit)} = \frac{\text{ml EDTA used} \times 1000}{\text{ml of sample}}$$

Alkalinity

Alkalinity was estimated by titrating the water sample with strong acid (HSO), first to pH 8.3, using phenolphthalein as an indicator and then to pH between 4.2 to 5.4 with methyl orange indicator.

100 ml of sample was taken in a conical flask and 2 drops of phenolphthalein indicator were added. Then, 2-3 drops of methyl orange were added.The colour of sample changed to yellow. The sample was titrated against 0.1 N HCl till yellow colour changed to pink, at the end point.

Alkalinity was calculated using the following formula:

$$\text{Total alkalinity (mg/lit)} = \frac{\text{ml} \times \text{N of HCl} \times 1000 \times 50}{\text{Volume of sample}}$$

pH

pH is the measure of the intensity of acidity or alkalinity and measures the concentration of hydrogen ions in water. pH was measured immediately after collecting the water. Systronics Digital pH meter was used for this purpose. The instrument was calibrated before each measurement by using buffer solution having pH 7.0. The electrode was rinsed with distilled water, dried with tissue paper and placed into the sample to be tested. The reading was noted down.

Chloride

Silver nitrate reacts with chloride to form very slightly soluble white precipitate of AgCl. At the end point, when all the chlorides get precipitated, free silver ions react with potassium chromate, to form silver chromate of red colour.

To determine the chloride content in water, 50 ml of sample was taken in a conical flask and 2 ml of potassium chromate indicator was added. The sample was titrated against 0.02 N silver nitrate solution until a brick end point appeared. Chloride was calculated by using the following formula:

$$\text{Chloride (mg/lit)} = \frac{(\text{ml} \times \text{N}) \text{ of Ag } NO_3 \times 1000 \times 35.5}{\text{ml of sample}}$$

Phosphates

The phosphorous, in the natural freshwater, is present mostly in inorganic forms such as $H_2PO_4^-$, HPO_4^{2-} and PO_4^{3-}.

Phosphates, in water, react with ammonium molybdate and form complex heteropoly acid (molybdophosphoric acid) which gets reduced to a complex of blue colour, in the presence of SnCl. The absorption of light by this blue colour can be measured at 690 nm to calculate the concentration of phosphates.

To determine the phosphates, 50 ml of filtered clear water sample was taken in a conical flask. 2 ml of ammonium molybdate, followed by 5 drops of stannous chloride solution were added. A blue colour developed after some time. After 5 minutes, but before 12 minutes, the readings were taken at 690 nm on a spectrophotometer, using distilled water as blank. The concentration was determined with the help of the standard curve.

Nitrates

Nitrates were determined by using *Brawne method*. Nitrate and brucine react to produce a yellow colour, the intensity of which can be measured at 410 nm.

10 ml of water sample was taken in a tube and 2 ml of NaCl solution was added. Contents were mixed properly and then 10 ml H_2SO_4 solution was added. 0.5 ml brucine sulphanilic acid solution was added and mixed thoroughly. The tube was placed in water bath for 15-20 minutes. After

20 minutes the tube was cooled and reading was taken at 410 nm using spectrophotometer. Concentration of nitrate was determined using standard curve.

Biological Parameters

Planktons

For biological study, samples were collected by filtering 50 liter surface water through a plankton net, made up of bolting silk cloth no. 20. Extreme care was taken in order to keep water undisturbed at the time of sampling. The collected samples were preserved in lugol's solution for phytoplanktons and in 4 per cent formalin for zooplanktons. The preserved samples were brought to the laboratory for qualitative and quantitative analysis. Phytoplanktons were identified by using the standard methods suggested by Smith (1950), Phillipose (1970) and Adoni (1985). Zooplanktons were identified by using the methods of Pennak (1978), Battish (1992), Dhanapati (2000) and Goswami (2000).

Quantitative studies were made by using Sedgwick rafter cell. Sample was properly agitated to distribute the organisms evenly. By using a pipette, one ml of sample was transferred onto the cell. Cover slip was placed properly avoiding any air bubble. Planktons were allowed to settle for some time and counting was made under microscope. All the planktons, present in the cell, were counted by moving the cell vertically and horizontally, covering the whole area.

Primary Productivity

Primary productivity was measured by using *Gaarder and Gran's method* of light and dark bottle. Two BOD bottles - one light and one dark painted bottle were filled with water and were suspended at same depth for 3 hours. Initial concentration of oxygen in the water was estimated separately by filling one more bottle (initial bottle). After 3 hours, the bottles were removed and their oxygen contents were measured. Primary productivity was measured by using the following formulae:

$$\text{Gross Primary Productivity (GPP) } (O_2 \text{ mg/lit/hr}) = \frac{Dl - Di}{h}$$

$$\text{Net Primary Productivity (NPP) } (O_2 \text{ mg/lit/hr}) = \frac{Dl - Dd}{h}$$

$$\text{Community Respiration (CP) } (O_2\text{mg/lit/hr}) = \frac{Di - Dd}{h}$$

where,

Di = Dissolved oxygen in the initial bottle (mg/lit)

Dl = Dissolved oxygen in the light bottle (mg/lit)

Dd = Dissolved oxygen in the dark bottle (mg/lit)

H = Duration of exposure in hours

The values were converted to carbon by using the following formula:

GPP/NPP/CP (in gC/m^3/hr) = GPP/NPP/CP (in mg O_2/lit/hr) × 0.375

5

Observations

The water quality of a water body depends on its physico-chemical characteristics. Physico-chemical factors influence the aquatic flora and fauna, which in turn, also influence the physico-chemical properties of water.

In addition, the climatic conditions of the area and its geographical location also influence the physico-chemical and biological properties of water. Thus, the quality of the water of a water body depends on several factors. To assess the water quality of a water body, determination of physico-chemical, biological and climatic conditions all are important.

Climatic Parameters (Table 3)

To determine climatic parameters - air, temperature, rainfall and humidity for each month, data were obtained from Metrological Department, Gwalior.

Air Temperature (Figure 5)

Monthly average air temperature (minimum and maximum temperature) were obtained from the Metrological Department for the study period from November 2010 to October 2011.The average minimum temperature varied from 6.37 to 27.71 °C. The lowest temperature was recorded during January 2011 while the highest average minimum temperature was recorded in May 2011.

Table 3: Climatic Parameters (Average Values for the Month) of Air Temperature, Rainfall and Humidity of Gwalior for November 2010 to October 2011*

Months	*Temperature (°C)*		*Rainfall (in mm)*	*Humidity*	
	Minimum	*Maximum*		*at 8.30 AM*	*at 5.30 PM*
Nov. 10	16.98	27.9	83.7	87.46	69.09
Dec. 10	8.86	24.24	43	90.06	60.51
Jan. 11	6.37	21.52	0	90.29	54.87
Feb. 11	12.04	26.56	7.4	85.36	47.71
Mar. 11	16.42	33.32	0	63.61	36
Apr. 11	20.86	37.81	9.6	50.8	30.7
May 11	27.71	42.55	20.9	42.45	28.93
Jun. 11	27.1	37.84	207.8	64.46	54.46
Jul. 11	26.56	32.81	340.5	81.9	70.32
Aug. 11	26.22	33.39	210.9	85.13	73.87
Sep. 11	24.47	32.44	180.1	76.33	64.5
Oct. 11	19.07	34.53	0	68	51.61
Minimum	6.37	21.52	0	42.45	28.93
Maximum	27.71	42.55	340.5	90.29	73.87

* Data obtained from *Metrological Department, Gwalior.*

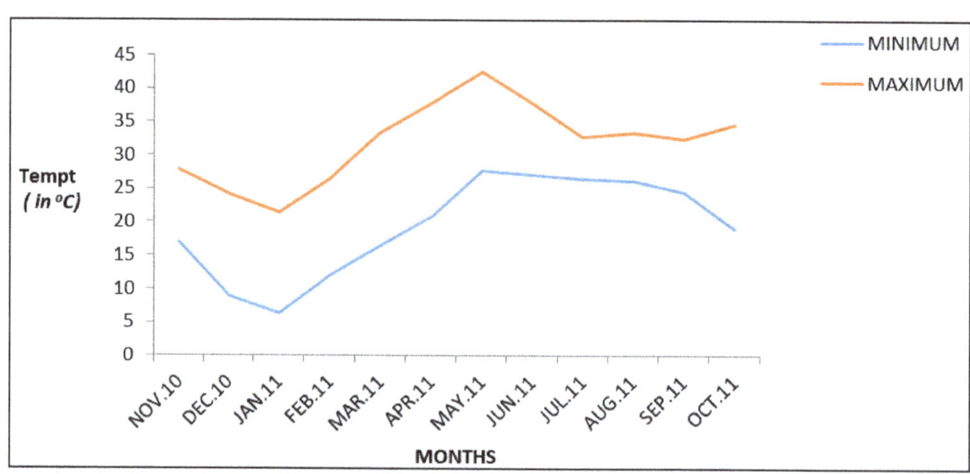

Figure 5: Climatic Parameters (Average value for the month) of Temperature of the Gwalior District for November 2010 to October 2011.

The average maximum temperature of the month varied from 21.52 °C to 42.55 °C. The lowest (21.52 °C) temperature was recorded during January 2011 and the highest temperature (42.55°C) was recorded during May 2011.

Minimum and maximum temperatures were higher during summer than monsoon and winter. The temperature was low during winter season.

Rainfall (Figure 6)

Total monthly rainfall ranged between 0 (nil) and 340.5 mm. Maximum rainfall was recorded during July, followed by August and June. There was no rainfall during October, January and March. Overall, a good rainfall was recorded during the study period.

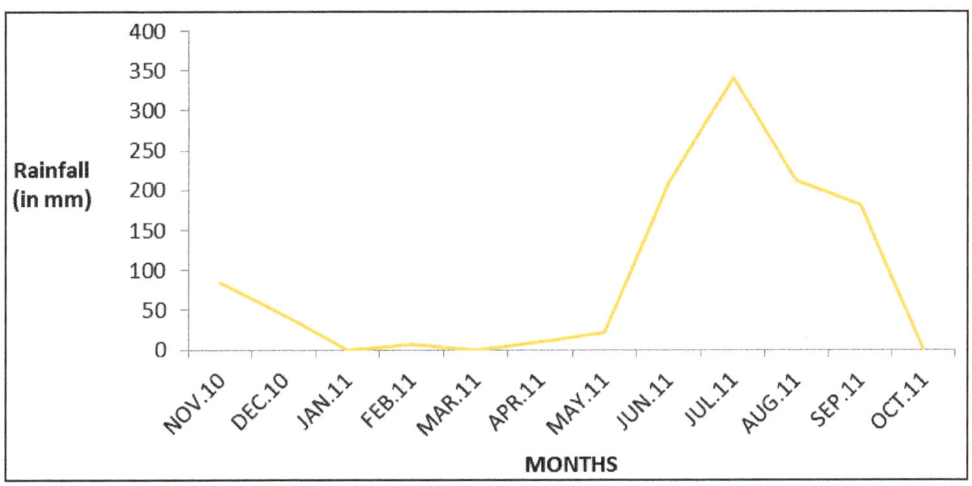

Figure 6: Climatic Parameters (Average value for the month) of Rainfall of the Gwalior District for November 2010 to October 2011.

Humidity (Figure 7)

Humidity was recorded at 8.30 AM and 5.30 PM daily by the Metrological Department. Data of the average humidity for the months were noted down. Average monthly humidity at 8.30 AM, which was the time close to the time of collection of the water samples, ranged between 42.45 per cent to 90.29 per cent. The average maximum humidity 90.29 per cent was recorded during January while humidity was minimum (42.45 per cent) during May. Humidity was higher during winter and low during summer. During August and September (monsoon), humidity was recorded as 85.13 per cent and 76.33 per cent, respectively.

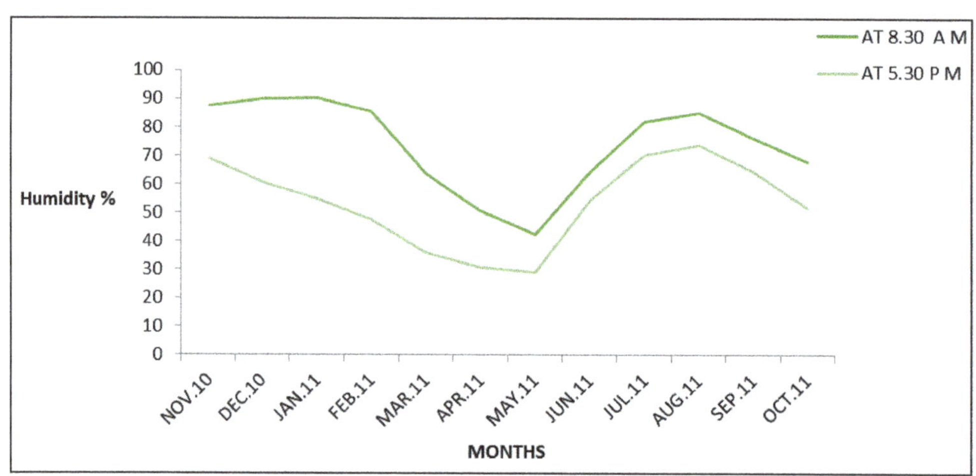

Figure 7: Climatic Parameters (Average value for the month) of Humidity of the Gwalior District for November 2010 to October 2011.

The average humidity recorded at 5.30 PM ranged between 28.93 per cent to 73.87 per cent. The maximum humidity was recorded during August while it was minimum during May.

Physical Parameters

Colour (Table 4)

The colour of the water of Tighra reservoir was found to be more or less transparent. During most of the study period it was transparent. During monsoon, the water was turbid. The colour of the water was in the range of 5 to 10 colour units. During most of the times it was transparent and of 5 units. The colour was under the standard range. No visible particle or colour was observed.

Water Temperature (Table 5, Figure 8)

The average water temperature varied from 18.4 °C to 35.75 °C during the study period. The average minimum temperature 18.4 °C was recorded in the month of January 2011 and the average maximum temperature was recorded in the month of June 2011. The minimum temperature 18 °C was recorded at Station 4 in January 2011, while the maximum temperature 36.2 °C was recorded at Station 2 in June 2011.

A gradual increase in water temperature was recorded from January to June 2011 that decreased from July to October 2011.

Table 4: Monthly Variations in Colour (in colour units) of Water at Four Stations of Tighra Reservoir, Gwalior for November 2010 to October 2011

Month	*November 2010 to October 2011*				*Mean*
	Stations				
	1	*2*	*3*	*4*	
Nov. 10	5	5	5	5	5
Dec. 10	5	5	5	5	5
Jan. 11	5	5	5	5	5
Feb. 11	5	5	5	5	5
Mar. 11	5	5	5	5	5
Apr. 11	5	5	5	5	5
May 11	5	5	5	5	5
Jun. 11	10	5	5	5	6.25
Jul. 11	10	5	10	10	8.75
Aug. 11	10	5	5	10	7.5
Sep. 11	5	5	5	5	5
Oct. 11	5	5	5	5	5
Minimum	5	5	5	5	5
Maximum	10	10	10	10	10
Summer	6.25	5	5	5	5.3125
Monsoon	7.5	5	6.25	7.5	6.5625
Winter	5	5	5	5	5

Conductivity (Table 6, Figure 9)

The conductivity of the water of Tighra reservoir was recorded in the range of 272.5 µS/cm to 408.5 µS/cm. It was highest during monsoon (360.5 µS/cm) and was lowest during winter (285.68 µS/cm). Maximum conductivity (415 µS/cm) was recorded at Station 2 during August 2011 and minimum conductivity of water (270 µS/cm) was recorded at Station 1 during December 2010.

Transparency (Table 7, Figure 10)

Transparency of the water of Tighra reservoir was in the range of 152.75 cm to 211.5 cm. Water was more transparent during winter (208.375cm) than summer (171.5 cm) and monsoon (161.31cm). Maximum transparency (217 cm) was recorded at Station 4 during December 2011and it was minimum

Table 5: Monthly Variations in Temperature (ºC) of Water at Four Stations of Tighra Reservoir, Gwalior for November 2010 to October 2011

Month	*November 2010 to October 2011*				*Mean*
	Stations				
	1	*2*	*3*	*4*	
Nov. 10	26	27.1	26.5	27	26.65
Dec. 10	20.2	20.2	20.1	20	20.125
Jan. 11	18.5	18.6	18.5	18	18.4
Feb. 11	20.6	20.5	20	20.2	20.325
Mar. 11	22	22.2	22.1	22.2	22.125
Apr. 11	26	25.7	25.9	25.2	25.7
May 11	32.8	32	32.4	32.2	32.35
Jun. 11	36	36.2	35.8	35	35.75
Jul. 11	32	32.4	32	32	32.1
Aug. 11	27	27.2	26.8	27	27
Sep. 11	25	25.2	25	25.2	25.1
Oct. 11	23.5	24	23.5	24	23.75
Minimum	18.5	18.6	18.5	18	18.4
Maximum	36	36.2	35.8	35	35.75
Summer	29.2	29.025	29.05	28.65	28.98125
Monsoon	26.875	27.2	26.825	27.05	26.9875
Winter	21.325	21.6	21.275	21.3	21.375

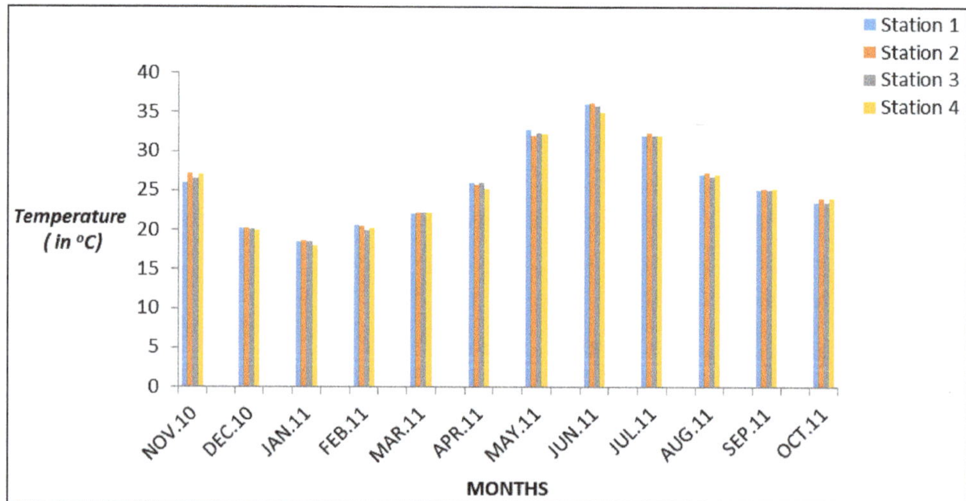

Figure 8: Monthly Variations in Temperature (in °C) of Water at Four Stations of Tighra Reservoir, Gwalior for November 2010 to October 2011.

Table 6: Monthly Variations in Conductivity (μS/cm) of Water at Four Stations of Tighra Reservoir, Gwalior for November 2010 to October 2011

Month	*November 2010 to October 2011*				*Mean*
	Stations				
	1	*2*	*3*	*4*	
Nov. 10	280	282	281	285	282
Dec. 10	270	271	275	274	272.5
Jan. 11	292	290	290	295	291.75
Feb. 11	299	295	292	300	296.5
Mar. 11	300	301	301	299	300.25
Apr. 11	305	304	302	307	304.5
May 11	295	299	300	302	299
Jun. 11	310	309	312	315	311.5
Jul. 11	360	370	369	372	367.75
Aug. 11	400	415	410	409	408.5
Sep. 11	340	350	342	345	344.25
Oct. 11	360	310	307	309	321.5
Minimum	270	271	275	274	272.5
Maximum	400	415	410	409	408.5
Summer	302.5	303.25	303.75	305.75	303.8125
Monsoon	365	361.25	357	358.75	360.5
Winter	285.25	284.5	284.5	288.5	285.6875

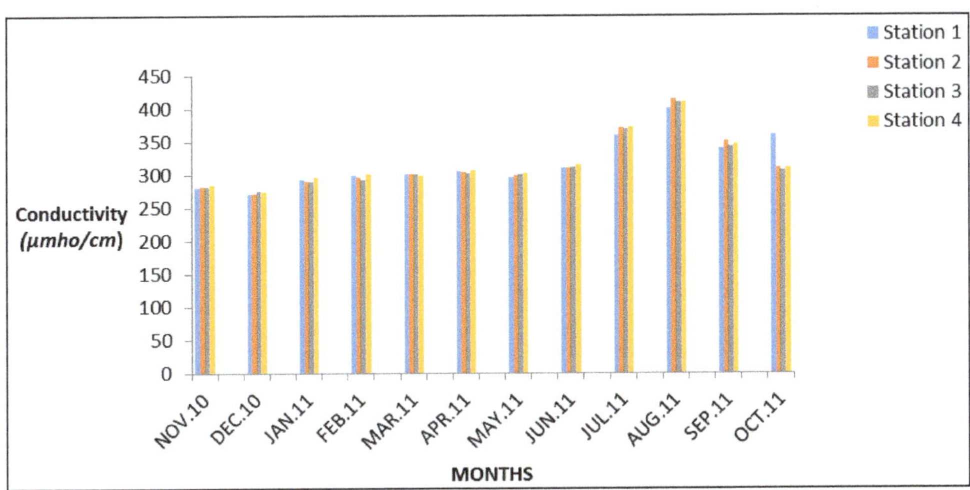

Figure 9: Monthly Variations in Conductivity (μS/cm) of Water at Four Stations of Tighra Reservoir, Gwalior for November 2010 to October 2011.

Table 7: Monthly Variations in Transparency (cm) of Water at Four Stations of Tighra Reservoir, Gwalior for November 2010 to October 2011

Month	*November 2010 to October 2011*				*Mean*
	Stations				
	1	*2*	*3*	*4*	
Nov. 10	195	215	215	210	208.75
Dec. 10	202	215	212	217	211.5
Jan. 11	210	202	208	210	207.5
Feb. 11	200	210	208	205	205.75
Mar. 11	190	195	190	185	190
Apr. 11	185	180	182	180	181.75
May 11	160	165	160	160	161.25
Jun. 11	155	152	150	155	153
Jul. 11	155	160	155	155	156.25
Aug. 11	150	159	152	154	153.75
Sep. 11	156	155	150	150	152.75
Oct. 11	185	175	170	200	182.5
Minimum	150	152	150	150	152.75
Maximum	210	215	215	217	211.5
Summer	172.5	173	170.5	170	171.5
Monsoon	161.5	162.25	156.75	164.75	161.3125
Winter	201.75	210.5	210.75	210.5	208.375

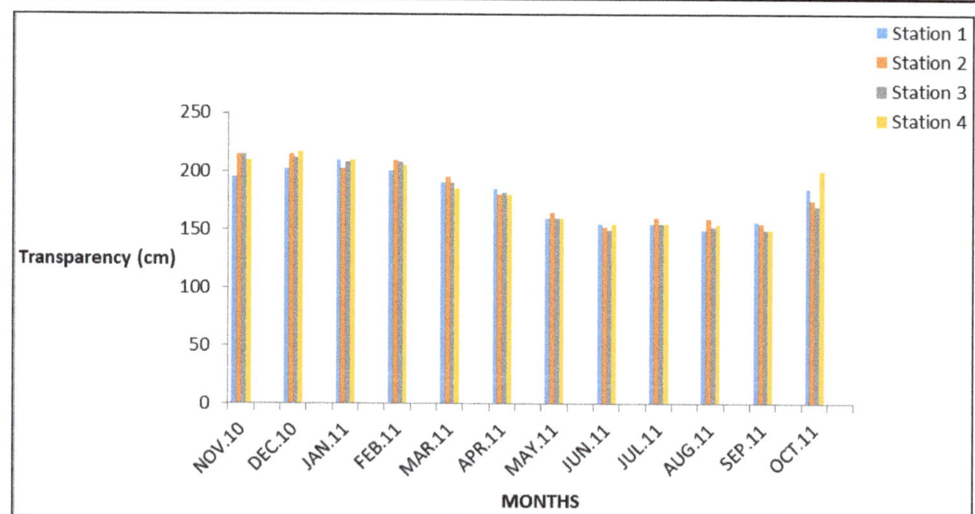

Figure 10: Monthly Variations in Transparency (cm) of Water at Four Stations of Tighra Reservoir, Gwalior for November 2010 to October 2011.

(150cm) during August 2011at Station 1 and September 2011 at Stations 3 and 4.

Turbidity (Table 8, Figure 11)

Turbidity of Tighra reservoir fell in the range of 5.77 NTU to 12.15 NTU. The water was more turbid during monsoon (9.95 NTU) than summer (8.67 NTU) and winter (6.21 NTU). Maximum turbidity (12.4 NTU) was recorded at Station 3 during August 2011 and minimum (5.6 NTU) was recorded at Station 2 during December 2010.

Chemical Parameters

pH (Table 9, Figure 12)

The average pH of the Tighra reservoir varied from 6.85 to 7.72. The minimum pH (6.6) was recorded at Station 4 in November 2010 and the maximum pH (8) was recorded at Station 1 and Station 4. The pH values show the water of the reservoir to be close to neutral, though it was slightly acidic during November 2010 and slightly alkaline during June 2011.

Dissolved Oxygen (DO) (Table10, Figure 13)

Dissolved oxygen of the reservoir was in the range of 5.425 to 8.125 mg/lit. The minimum value 5.4 mg/lit was recorded at Station 3 during June 2011, while the maximum value (8.2 mg/lit) was recorded at Stations 1 and 3, during December 2010, January 2011 and at Station 4 (December 2010, respectively).

Seasonally, the average minimum value (6.16 mg/lit) was recorded during summer and maximum dissolved oxygen (7.975 mg/lit) was recorded during winter.

Free CO_2 (Table 11, Figure 14)

Free CO_2 was recorded between 4.15 mg/lit and 7.57 mg/lit. It was low during winter (average 4.55 mg/lit), which increased during summer (5.98 mg/lit) and was maximum (6.556 mg/lit) during monsoon. The maximum free CO_2 (8 mg/lit) was recorded at Station 1 during September 2011 and at Station 3 during June 2011.

Table 8: Monthly Variations in Turbidity (NTU) of Water at Four Stations of Tighra Reservoir, Gwalior for November 2010 to October 2011

Month	*November 2010 to October 2011*				*Mean*
	Stations				
	1	*2*	*3*	*4*	
Nov. 10	6	6.1	5.9	6.2	6.05
Dec. 10	5.7	5.6	5.9	5.9	5.775
Jan. 11	6.5	6.8	6.7	6.7	6.675
Feb. 11	6.4	5.9	6.5	6.6	6.35
Mar. 11	7	7.2	7.1	7.1	7.1
Apr. 11	8.1	8.3	8.2	8.1	8.175
May 11	9	9.2	9.1	9.4	9.175
Jun. 11	10	10.2	10.3	10.5	10.25
Jul. 11	11.2	11.3	11	11	11.125
Aug. 11	12.1	12	12.4	12.1	12.15
Sep. 11	10.8	10.5	10.6	10.5	10.6
Oct. 11	6	5.8	5.9	6	5.925
Minimum	5.7	5.6	5.9	5.9	5.775
Maximum	12.1	12	12.4	12.1	12.15
Summer	8.525	8.725	8.675	8.775	8.675
Monsoon	10.025	9.9	9.975	9.9	9.95
Winter	6.15	6.1	6.25	6.35	6.2125

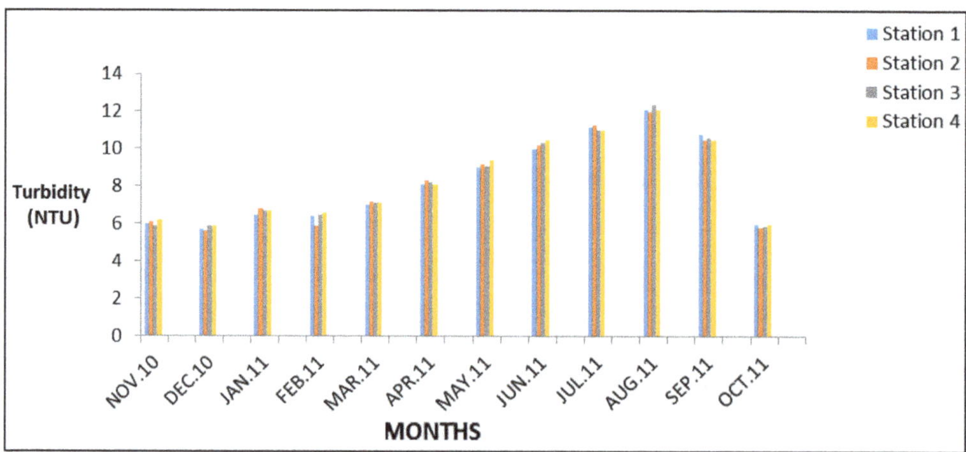

Figure 11: Monthly Variations in Turbidity (NTU) of Water at Four Stations of Tighra Reservoir, Gwalior for November 2010 to October 2011.

Table 9: Monthly Variations in pH of Water at Four Stations of Tighra Reservoir, Gwalior for November 2010 to October 2011

Month	*November 2010 to October 2011*				*Mean*
	Stations				
	1	*2*	*3*	*4*	
Nov. 10	7.1	6.9	6.8	6.6	6.85
Dec. 10	7.4	7.1	7	6.9	7.1
Jan. 11	6.9	6.8	7	7.2	6.975
Feb. 11	7.2	7.4	7.4	7.7	7.425
Mar. 11	7.9	7.7	7.4	7.7	7.675
Apr. 11	7.5	7.4	7.6	7.4	7.475
May 11	7.4	7.5	7.5	7.5	7.475
Jun. 11	8	7.1	7.8	8	7.725
Jul. 11	7.6	7.4	7.4	7.6	7.5
Aug. 11	7.6	7.2	7.4	7.6	7.45
Sep. 11	7.2	7.1	7.3	7.3	7.225
Oct. 11	7.2	7.1	7.3	7.4	7.25
Minimum	6.9	6.8	6.8	6.6	6.85
Maximum	8	7.7	7.8	8	7.725
Summer	7.7	7.425	7.575	7.65	7.5875
Monsoon	7.4	7.2	7.35	7.475	7.35625
Winter	7.15	7.05	7.05	7.1	7.0875

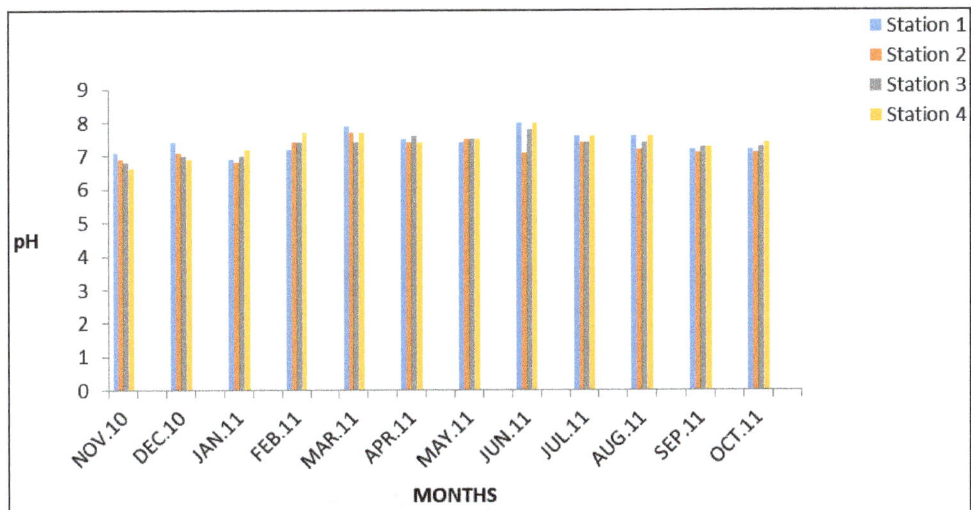

Figure 12: Monthly Variations in pH of Water at Four Stations of Tighra Reservoir, Gwalior for November 2010 to October 2011.

Table 10: Monthly Variations in Dissolved Oxygen (mg/l) of Water at Four Stations of Tighra Reservoir, Gwalior for November 2010 to October 2011

Month	*November 2010 to October 2011*				*Mean*
	Stations				
	1	*2*	*3*	*4*	
Nov. 10	8	8.1	8.2	7.9	8.05
Dec. 10	8.2	8.1	8	8.2	8.125
Jan. 11	8.1	8	8.2	7.9	8.05
Feb. 11	7.7	7.7	7.6	7.7	7.675
Mar. 11	7	7	7.1	7.2	7.075
Apr. 11	6.6	6.5	6.4	6.7	6.55
May 11	5.6	5.5	5.6	5.8	5.625
Jun. 11	5.5	5.2	5.4	5.6	5.425
Jul. 11	6.5	6.2	6.4	6.4	6.375
Aug. 11	7	7	6.9	7.1	7
Sep. 11	7.5	7.1	7.2	7	7.2
Oct. 11	7.8	7.7	7	7.9	7.6
Minimum	5.6	5.5	5.4	5.6	5.425
Maximum	8.2	8.1	8.2	8.2	8.125
Summer	6.175	6.05	6.125	6.325	6.16875
Monsoon	7.2	7	6.875	7.1	7.04375
Winter	8	7.975	8	7.925	7.975

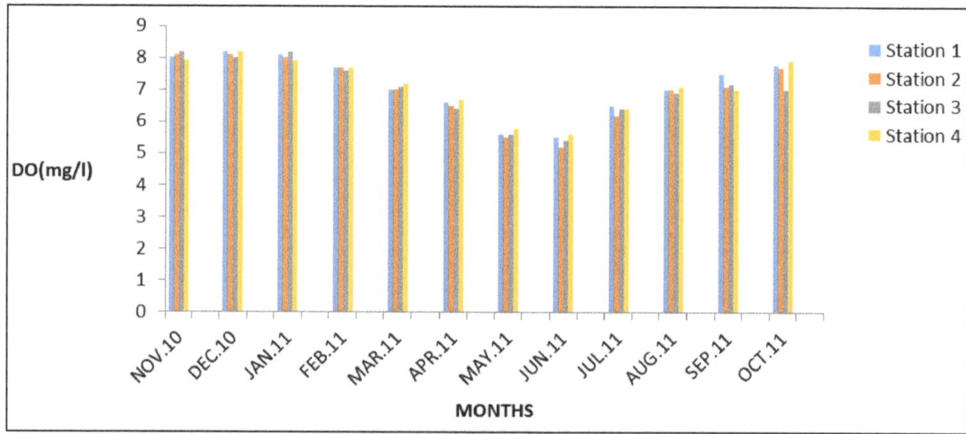

Figure 13: Monthly Variations in Dissolved Oxygen (mg/l) of Water at Four Stations of Tighra Reservoir, Gwalior for November 2010 to October 2011.

Table 11: Monthly Variations in Free CO_2 (mg/l) of Water at Four Stations of Tighra Reservoir, Gwalior for November 2010 to October 2011

Month	*November 2010 to October 2011*				*Mean*
	Stations				
	1	*2*	*3*	*4*	
Nov. 10	4.6	4.1	3.9	4	4.15
Dec. 10	5	4.7	6.1	5.2	5.25
Jan. 11	5.6	4.2	4.1	4.7	4.65
Feb. 11	4.2	4.5	3.9	4	4.15
Mar. 11	6	5.4	4.2	4.2	4.95
Apr. 11	6	4.1	4.2	4	4.575
May 11	7	6.9	6.5	7.1	6.875
Jun. 11	6.9	7.8	8	7.5	7.55
Jul. 11	7	7.8	8	7.5	7.575
Aug. 11	6.1	6	6.6	6.2	6.225
Sep. 11	8	6.5	7	6.9	7.1
Oct. 11	6.2	5.2	4.9	5	5.325
Minimum	4.2	4.1	3.9	4	4.15
Maximum	8	7.8	8	7.5	7.57
Summer	6.475	6.05	5.725	5.7	5.9875
Monsoon	6.825	6.375	6.625	6.4	6.55625
Winter	4.85	4.375	4.5	4.475	4.55

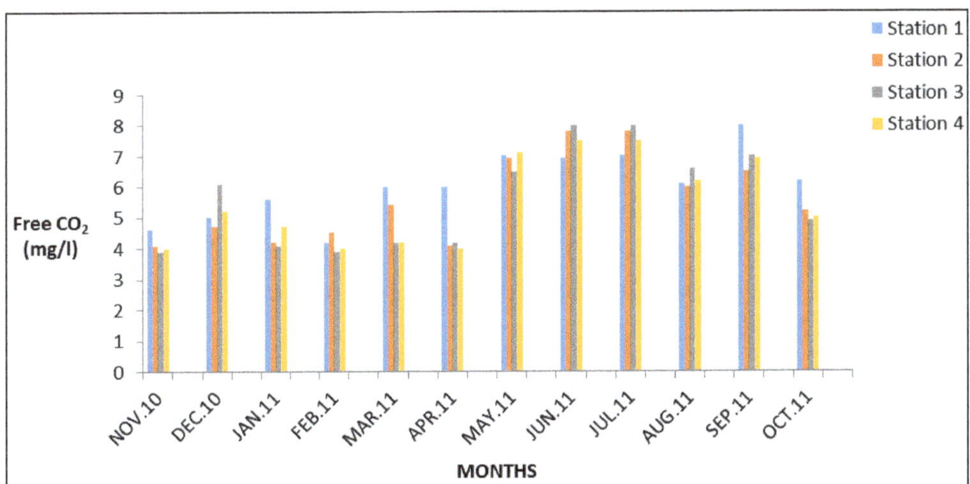

Figure 14: Monthly Variations in Free CO_2 (mg/l) of Water at Four Stations of Tighra Reservoir, Gwalior for November 2010 to October 2011.

Total Hardness (Table 12, Figure 15)

Total hardness of Tighra reservoir was recorded between 66.25 mg/lit and 137 mg/lit. Total hardness was highest during summer season (116.68 mg/lit), followed by monsoon (95.81 mg/lit) and minimum during winter (73.93 mg/lit). The maximum hardness (141 mg/lit) was recorded at Station 1 during June 2011while minimum (65 mg/lit) was recorded at Stations 1 and 2 during December 2010.

Alkalinity (Table 13, Figure 16)

Water of Tighra reservoir showed alkalinity in the range of 53.75 mg/lit to 145.5 mg/lit). Alkalinity was lowest (68.5 mg/lit) during monsoon, increased during winter (82.125 mg/lit) and was highest (123.56 mg/lit) during summer. The maximum alkalinity (160 mg/lit) was recorded during May 2011 at Station 1 and minimum alkalinity (48 mg/lit) was recorded at Station 2 during November 2010.

Chloride (Table 14, Figure 17)

Chloride varied from 11.85 mg/lit to 39.5 mg/lit.. Minimum chloride content was recorded at Station 4 (11.3mg/lt) during September 2011, while maximum chloride content was recorded at Station 2, during May 2011, which was 40.7 mg/lit.

Seasonally, chloride was maximum (36.81 mg/lit) during summer, followed by winter (21.39 mg/lit) and minimum (19.21 mg/lit) during monsoon.

Nitrates (Table 15, Figure 18)

Nitrate content,in Tighra reservoir, was low throughout the study. The values varied from 0.29 mg/lit to 0.61 mg/lit. Highest value (0.9 mg/lit) was recorded at Station 2 and 3 during April 2011 and October 2011, respectively. Lowest value 0.11 mg/lit was recorded at Station 1 during October 2011. Seasonally no considerable change in nitrate content of the water was observed.

Phosphate (Table 16, Figure 19)

The phosphate content of the Tighra reservoir was found in the range of 0.37 mg/lit to 1.57 mg/lit. The maximum value (1.7 mg/lit)was recorded

Table 12: Monthly Variations in Total Hardness (mg/l) of Water at Four Stations of Tighra Reservoir, Gwalior for November 2010 to October 2011

Month	*November 2010 to October 2011*				*Mean*
	Stations				
	1	*2*	*3*	*4*	
Nov. 10	70	72	68	69	69.75
Dec. 10	65	65	66	69	66.25
Jan. 11	67	69	70	71	69.25
Feb. 11	90	92	91	89	90.5
Mar. 11	89	85	90	88	88
Apr. 11	110	101	107	109	106.75
May 11	140	135	130	135	135
Jun. 11	141	140	132	135	137
Jul. 11	110	109	111	108	109.5
Aug. 11	95	94	97	96	95.5
Sep. 11	90	91	89	92	90.5
Oct. 11	90	89	87	85	87.75
Minimum	65	65	66	69	66.25
Maximum	141	140	132	135	137
Summer	120	115.25	114.75	116.75	116.6875
Monsoon	96.25	95.75	96	95.25	95.8125
Winter	73	74.5	73.75	74.5	73.9375

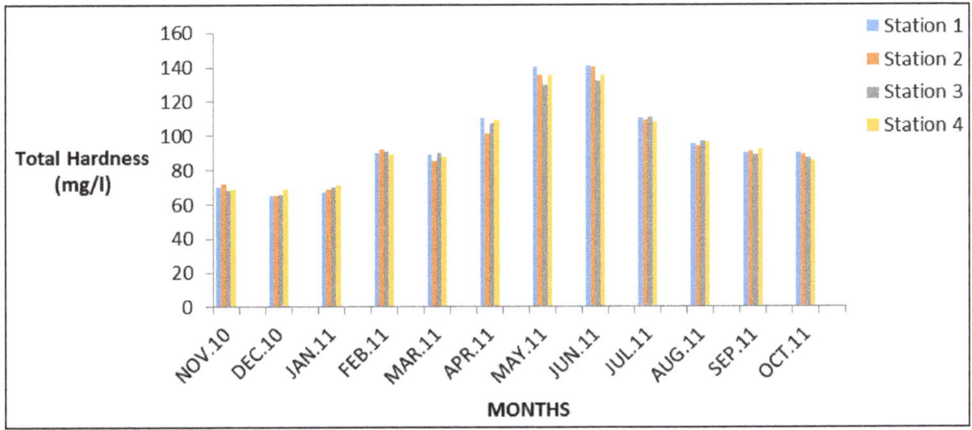

Figure 15: Monthly Variations in Total Hardness (mg/l) of Water at Four Stations of Tighra Reservoir, Gwalior for November 2010 to October 2011.

Table 13: Monthly Variations in Alkalinity (mg/l) of Water at Four Stations of Tighra Reservoir, Gwalior for November 2010 to October 2011

Month	*November 2010 to October 2011*				*Mean*
	Stations				
	1	*2*	*3*	*4*	
Nov. 10	52	48	55	60	53.75
Dec. 10	90	91	91	85	89.25
Jan. 11	90	95	98	91	93.5
Feb. 11	98	90	88	92	92
Mar. 11	130	140	130	110	127.5
Apr. 11	120	110	115	102	111.75
May 11	160	150	152	120	145.5
Jun. 11	110	101	125	102	109.5
Jul. 11	85	84	80	90	84.75
Aug. 11	50	49	60	72	57.75
Sep. 11	72	60	71	67	67.5
Oct. 11	80	59	62	55	64
Minimum	50	48	55	55	53.75
Maximum	160	150	152	120	145.5
Summer	130	125.25	130.5	108.5	123.5625
Monsoon	71.75	63	68.25	71	68.5
Winter	82.5	81	83	82	82.125

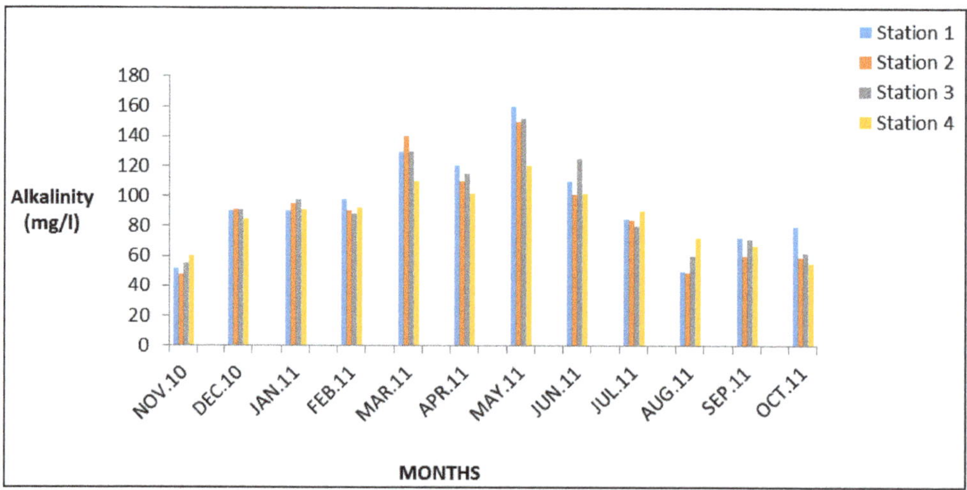

Figure 16: Monthly Variations in Alkalinity (mg/l) of Water at Four Stations of Tighra Reservoir, Gwalior for November 2010 to October 2011.

Table 14: Monthly Variations in Chloride (mg/l) of Water at Four Stations of Tighra Reservoir, Gwalior for November 2010 to October 2011

Month	*November 2010 to October 2011*				*Mean*
	Stations				
	1	*2*	*3*	*4*	
Nov. 10	14.5	20.6	21.3	16.2	18.15
Dec. 10	14.85	15.8	15.3	14.7	15.1625
Jan. 11	20.4	25.8	24.2	24	23.6
Feb. 11	28.9	29	29.3	27.4	28.65
Mar. 11	30.4	30.2	32.8	32.1	31.375
Apr. 11	39.2	38.6	37.2	37.2	38.05
May 11	36.7	40.7	39	36.9	38.325
Jun. 11	40.5	39.2	40.1	38.2	39.5
Jul. 11	36.4	37.1	35.1	32	35.15
Aug. 11	14.1	15.2	14.2	12.4	13.975
Sep. 11	12.1	11.5	12.5	11.3	11.85
Oct. 11	15.27	16.4	16.3	15.5	15.8675
Minimum	12.1	11.5	12.5	11.3	11.85
Maximum	40.5	40.7	40.1	38.2	39.5
Summer	36.7	37.175	37.275	36.1	36.8125
Monsoon	19.4675	20.05	19.525	17.8	19.21063
Winter	19.6625	22.8	22.525	20.575	21.39063

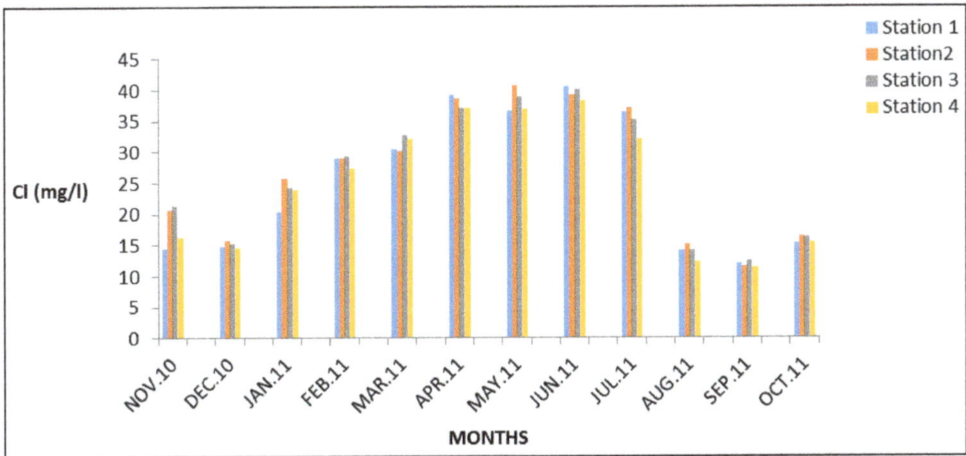

Figure 17: Monthly Variations in Chloride (mg/l) of Water at Four Stations of Tighra Reservoir, Gwalior for November 2010 to October 2011.

Table 15: Monthly Variations in Nitrates (mg/l) of Water at Four Stations of Tighra Reservoir, Gwalior for November 2010 to October 2011

Month	*November 2010 to October 2011*				*Mean*
	Stations				
	1	*2*	*3*	*4*	
Nov. 10	0.62	0.62	0.52	0.69	0.6125
Dec. 10	0.43	0.5	0.72	0.7	0.5875
Jan. 11	0.39	0.4	0.48	0.42	0.4225
Feb. 11	0.27	0.3	0.31	0.29	0.2925
Mar. 11	0.42	0.41	0.4	0.5	0.4325
Apr. 11	0.32	0.9	0.6	0.7	0.63
May 11	0.38	0.4	0.37	0.38	0.3825
Jun. 11	0.41	0.32	0.3	0.41	0.36
Jul. 11	0.5	0.6	0.61	0.49	0.55
Aug. 11	0.39	0.4	0.41	0.42	0.405
Sep. 11	0.28	0.21	0.3	0.87	0.415
Oct. 11	0.11	0.42	0.9	0.7	0.5325
Minimum	0.11	0.21	0.31	0.29	0.29
Maximum	0.62	0.9	0.9	0.87	0.61
Summer	0.3825	0.5075	0.4175	0.4975	0.45125
Monsoon	0.32	0.4075	0.555	0.62	0.475625
Winter	0.4275	0.455	0.5075	0.525	0.47875

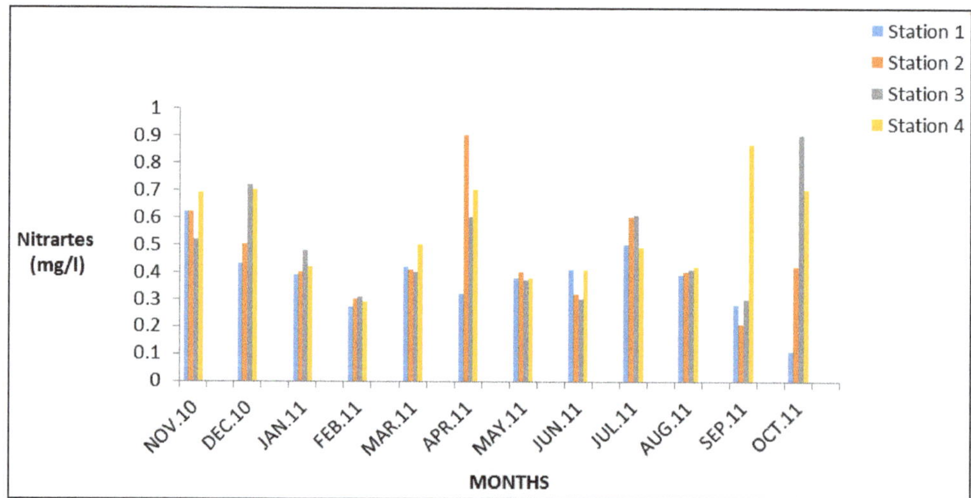

Figure 18: Monthly Variations in Nitrates (mg/l) of Water at Four Stations of Tighra Reservoir, Gwalior for November 2010 to October 2011.

Table 16: Monthly Variations in Phosphates (mg/l) of Water at Four Stations of Tighra Reservoir, Gwalior for November 2010 to October 2011

Month	*November 2010 to October 2011*				*Mean*
	Stations				
	1	*2*	*3*	*4*	
Nov. 10	1	1.2	1.1	1.4	1.175
Dec. 10	1.6	1.4	1.5	1.5	1.5
Jan. 11	1.4	1.2	1.4	1.4	1.35
Feb. 11	1.3	1.4	1.4	1.3	1.35
Mar. 11	1.5	1.6	1.5	1.7	1.575
Apr. 11	1.7	1.5	1.6	1.5	1.575
May 11	1.6	1.6	1.6	1.5	1.575
Jun. 11	1.4	1.3	1.4	1.4	1.375
Jul. 11	1.1	1	1.2	1	1.075
Aug. 11	0.4	0.4	0.3	0.4	0.375
Sep. 11	0.6	0.9	0.9	1	0.85
Oct. 11	0.2	0.8	0.7	0.8	0.625
Minimum	0.2	0.4	0.3	0.4	0.375
Maximum	1.7	1.6	1.6	1.7	1.575
Summer	1.55	1.5	1.525	1.525	1.525
Monsoon	0.575	0.775	0.775	0.8	0.73125
Winter	1.325	1.3	1.35	1.4	1.34375

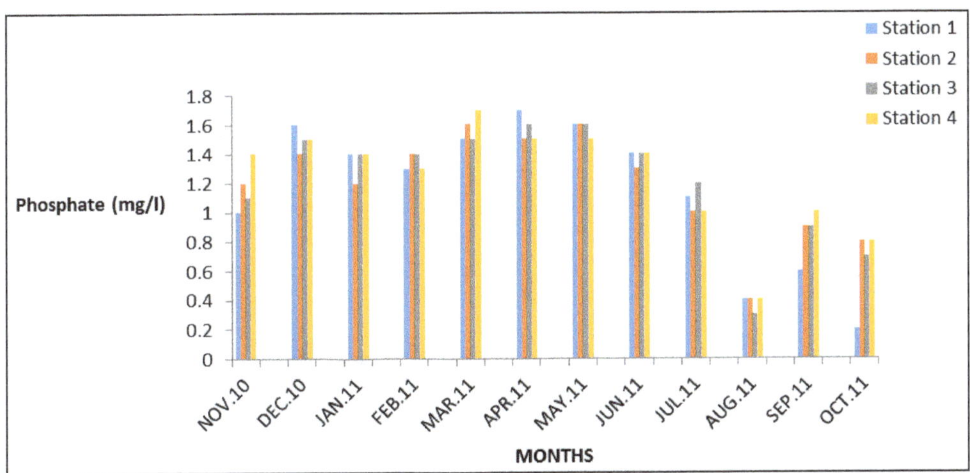

Figure 19: Monthly Variations in Phosphates (mg/l) of Water at Four Stations of Tighra Reservoir, Gwalior for November 2010 to October 2011.

at Stations 1 and 4 during April 2011 and March 2011, respectively. The minimum value (0.2 mg/lit) was recorded at Station 1 during October 2011.

Seasonally, phosphate was low during monsoon (0.731 mg/lit). During summer and winter, the phosphate content of the water was 1.52 mg/lit and 1.34 mg/lit, respectively.

Biological Studies

Planktons

Both phytoplanktons and zooplanktons were recorded during the study period.Total number of planktons, collected at all the four stations, were combined to get the average number of planktons collected during a month.

Phytoplanktons (Tables 17–20, Figures 20–24)

In Tighra reservoir all four groups of phytoplanktons, Bacillariophyceae, Chlorophyceae, Myxophyceae and Euglenophyceae were recorded. Bacillariophyceae was the most dominant group, representing 39.81 per cent of the total phytoplanktons, followed by Myxophyceae (29.21 per cent), Chlorophyceae (20.68 per cent) and Euglenophyceae (10.28 per cent).

Among the total 26 genera of phytoplanktons, observed in Tighra reservoir, 9 were from Bacillariophyceae, 9 from Chlorophyceae, 6 from Myxophyceae and 2 from Euglenophyceae. List of these phytoplanktons is mentioned in Table 18.

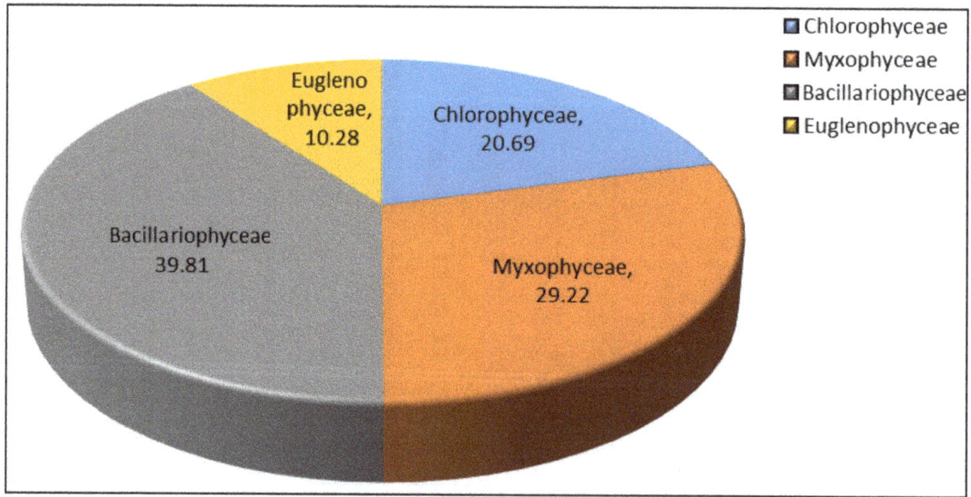

Figure 20: Annual Variations in Phytoplankton Composition at Tighra Reservoir, Gwalior for November 2010 to October 2011.

Table 17: Monthly Variations in Phytoplankton (No./lit) of Water at Four Stations of Tighra Reservoir, Gwalior for November 2010 to October 2011

	Nov	*Dec*	*Jan*	*Feb*	*Mar*	*Apr*	*May*	*Jun*	*Jul*	*Aug*	*Sep*	*Oct*	*TOTAL*
Chlorophyceae													
Station 1	168	235	412	310	360	415	310	320	355	275	290	300	3750
Station 2	170	202	300	300	265	315	265	310	400	210	210	250	3197
Station 3	175	215	455	275	250	415	300	275	390	255	200	190	3395
Station 4	160	155	390	300	300	390	275	290	300	200	165	155	3080
Mean	168.25	201.75	389.25	296.25	293.75	383.75	287.5	298.75	361.25	235	216.25	223.75	**13422**
Myxophyceae													
Station 1	328	355	375	390	422	580	556	465	445	402	310	318	4946
Station 2	365	352	360	400	405	572	565	412	430	352	366	365	4944
Station 3	375	375	380	375	412	566	505	455	402	312	315	290	4762
Station 4	255	265	310	312	410	515	502	410	400	333	300	290	4302
Mean	330.75	336.75	356.25	369.25	412.25	558.25	532	435.5	419.25	349.75	322.75	315.75	**18954**
Bacillariophyceae													
Station 1	425	400	265	525	625	952	810	772	675	415	465	482	6811
Station 2	420	390	300	550	590	1052	900	800	690	401	315	352	6760
Station 3	390	275	275	412	598	915	812	798	602	452	300	252	6081
Station 4	455	265	352	398	552	985	810	655	570	490	355	290	6177
Mean	422.5	332.5	298	471.25	591.25	976	833	756.25	634.25	439.5	358.75	344	**25829**

Contd...

Table 17–*Contd...*

	Nov	*Dec*	*Jan*	*Feb*	*Mar*	*Apr*	*May*	*Jun*	*Jul*	*Aug*	*Sep*	*Oct*	*TOTAL*
Euglenophyceae													
Station 1	175	125	215	205	238	312	120	115	98	90	90	75	1858
Station 2	112	155	165	210	210	285	100	110	90	75	90	100	1702
Station 3	165	110	190	190	175	255	110	200	115	90	65	85	1750
Station 4	110	100	100	175	110	210	90	170	75	65	90	65	1360
Mean	140.5	122.5	167.5	195	183.25	265.5	105	148.75	94.5	80	83.75	81.25	**6670**

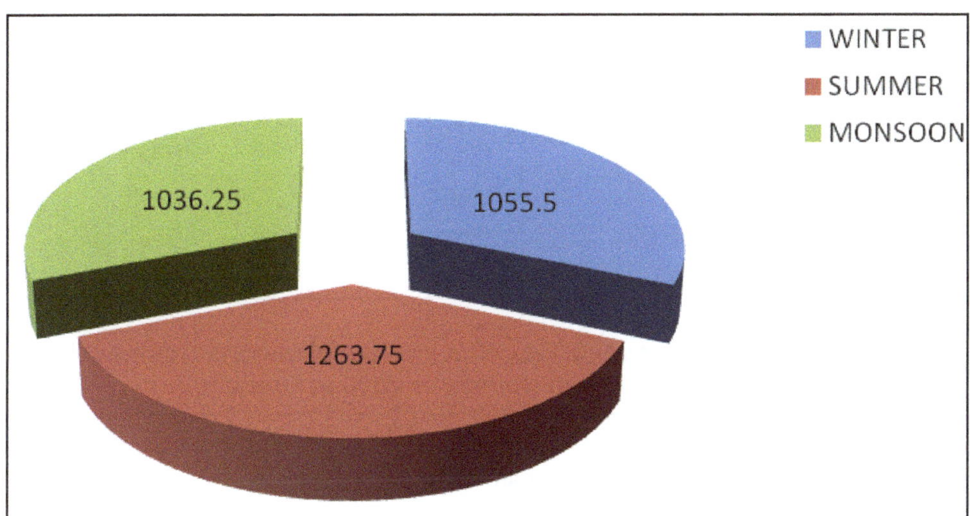

Figure 21: Seasonal Variations in Chlorophyceae (Cells/lit) at Tighra Reservoir, Gwalior for November 2010 to October 2011.

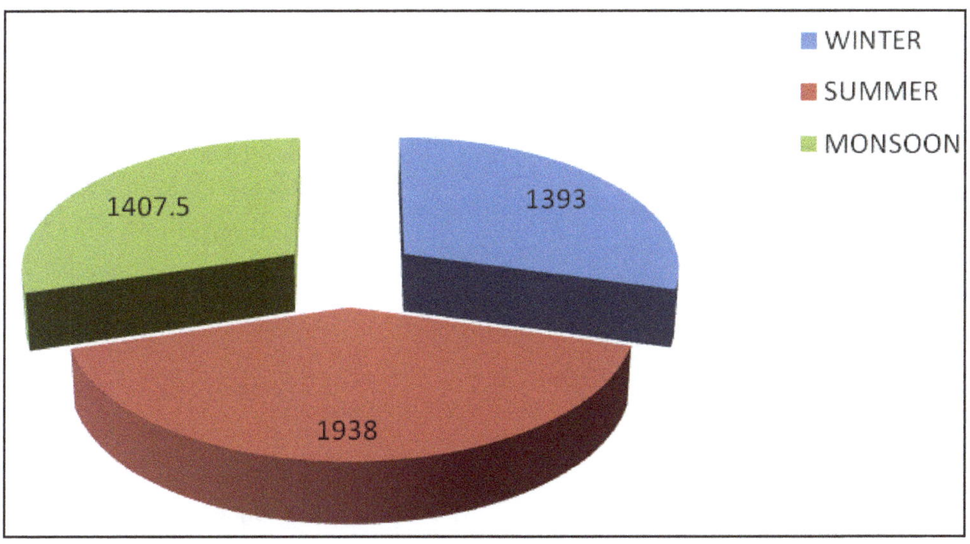

Figure 22: Seasonal Variations in Myxophyceae (Cells/lit) at Tighra Reservoir, Gwalior for November 2010 to October 2011.

Bacillariophyceae were represented by 9 genera, *Cyclotella operculata*, *Synedra* sp., *Navicula* sp., *Melosira* sp., *Diatoms* sp., *Fragilaria* sp., *Cymbella* sp., *Tabellaria* sp.and *Pinnularia viridis*. The number of bacillariophyceae were found to be highest during April 2011 (1052 cells/lit) at Station 2. The

Table 18: Annual Variation in Phytoplanktons Composition in Tighra Reservoir from November 2010 to October 2011

	Number of Organisms	*Percentage*
Chlorophyceae	13422	20.689
Myxophyceae	18954	29.216
Bacillorophyceae	25829	39.813
Euglenophyceae	6670	10.281

Table 19: Group-wise Seasonal Variation in Phytoplanktons Composition in Tighra Reservoir from November 2010 to October 2011

Season	*Chlorophyceae*	*Myxophyceae*	*Bacillariophyceae*	*Euglenophyceae*
Summer	1263.75	1938	3156.5	702.5
Monsoon	1036.25	1407.5	1776.5	339.5
Winter	1055.5	1393	1524.25	625.5

Table 20: List of Phytoplanktons Recorded in Tighra Reservoir from November 2010 to December 2011

BACILLARIOPHYCEAE	
Cyclotella operculata	*Synedra* sp.
Navicula sp.	*Melosira* sp.
Diatoms sp.	*Fragilaria* sp.
Cymbella sp.	*Tabellaria* sp.
Pinnularia viridis	
CHLOROPHYCEAE	
Pandorina morum	*Scenedesmus* sp.
Pediastrum duplex	*Ankistrodesmus* sp.
Chlorella vulgaris	*Closterium* sp.
Spirogyra sp.	*Oedogonium* sp.
Ulothrix sp.	
MYXOPHYCEAE	
Microcystis sp.	*Merismopedia* sp.
Oscillatoria chlorina	*Spirulina* sp.
Anabaena sp.	*Nostoc* sp.
EUGLENOPHYCEAE	
Phacus sp.	*Euglena* sp.

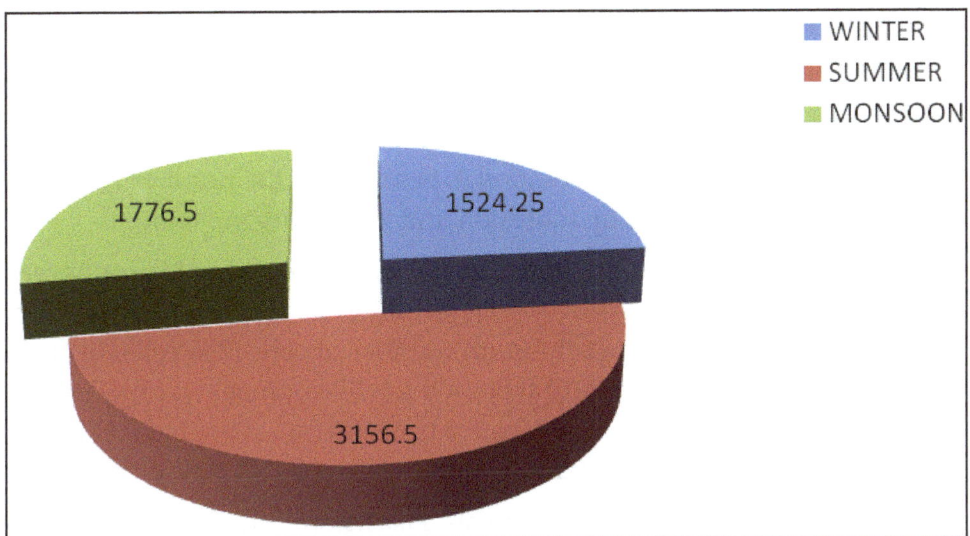

Figure 23: Seasonal Variations in Bacillariophyceae (Cells/lit) at Tighra Reservoir, Gwalior for November 2010 to October 2011.

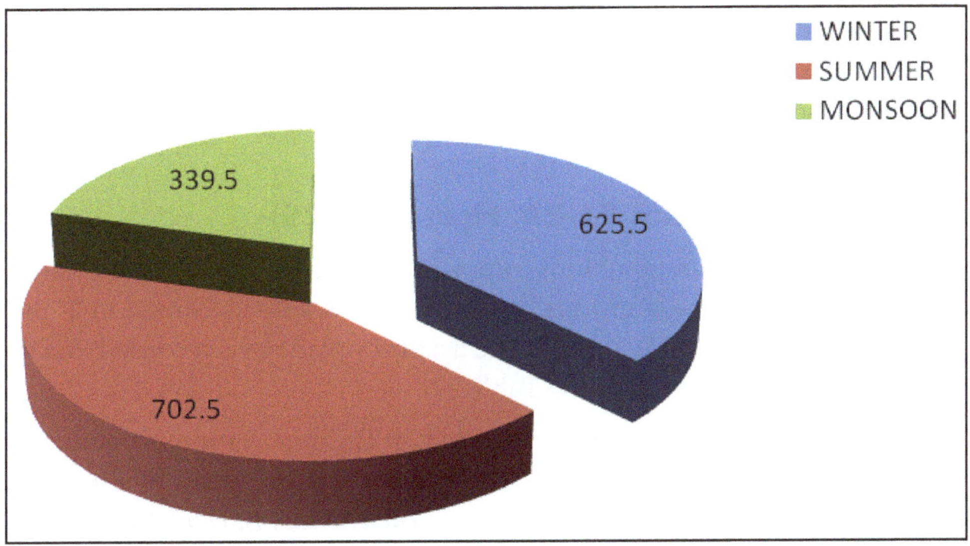

Figure 24: Seasonal Variations in Euglenophyceae (Cells/lit) at Tighra Reservoir, Gwalior for November 2010 to October 2011.

minimum number (252 cells/lit) was recorded during October 2010. The number of cells recorded was more during summer months.

9 genera of Chlorophyceae, identified from Tighra reservoir were – *Pandorina morum*, *Scenedesmus* sp., *Pediastrum duplex*, *Ankistrodesmus* sp.,

Chlorella vulgaris, Closterium sp., *Spirogyra* sp., *Oedogonium* sp. and *Ulothrix* sp. Highest number of Chlorophyceae were recorded during January 2011 (453 cells/lit) at Station 3, followed by 415 cells/lit during April 2011 at the same station. Minimum number of Chlorophyceae were recorded during October 2011 (155 cells/lit) at Station 4. Seasonally, the number was more during summer, followed by monsoon and lowest during winter.

Out of total 26 genera of Phytoplanktons, 6 belonged to Myxophyceae. They were *Microcystis* sp., *Merismopedia sp., Oscillatoria chlorina, Spirulina* sp., *Anabaena* sp. and *Nostoc* sp. Minimum number of cells (255 cells/lit) were recorded during November 2010 at Station 4, while number was highest (580 cells/lit) during April 2011at Station 1. *Microcystis* sp., *Merismopedia* sp.and *Spirulina* sp. were found in maximum numbers among all the Myxophyceae.

Only 2 genera of Euglenophyceae, *Phacus* sp. and *Euglena* sp. were identified. They represented 10.28 per cent of the total phytoplanktons. The number was maximum during April 2011 (265 cells/lit) and it was minimum during August 2011 (80 cells/lit). The number was lower during winter months, in comparison to summer months. The highest number (312 cells/lit) was recorded at Station 1 during April 2011 and minimum number (65 cells/lit) was recorded at Station 3 (September 2011) and Station 4 (August 2011 and October 2011).

Zooplanktons (Tables 21–24, Figures 25–29)

The freshwater zooplanktons comprise of Protozoa, rotifers, cladocerans and copepods. All these groups of zooplanktons wee identified in Tighra reservoir during the study period. The list of zooplanktons recorded, during the study period, is given in the Table 22.

Among all the zooplanktons recorded, in Tighra reservoir, rotifers were found to the most dominant group. The group was represented by *Brachionus angularis, B. quadridentatus, B. falcatus, B. calyciflorus, B. caudatus, Keratella tropica, Lepadella patella, Polyarthra* sp.,*Rotaria sp.* and *Filina longista.* Rotifers were found to be dominant throughout the study. Rotifers represented 39 per cent of all zooplanktons, recorded from the Tighra reservoir. The maximum number of rotifers (525 cells/lit) was recorded at Station 4 during May 2011 and the number was minimum(185 cells/lit) at Station 4 in December 2010. Seasonally, the number was highest during summer, followed by monsoon and lowest during winter.

Table 21: Monthly Variations in Zooplanktons (No./lit) of Water at Four Stations of Tighra Reservoir, Gwalior for November 2010 to October 2011

	Nov	*Dec*	*Jan*	*Feb*	*Mar*	*Apr*	*May*	*Jun*	*Jul*	*Aug*	*Sep*	*Oct*	*TOTAL*
Rotifera													
Station 1	180	190	220	240	300	420	475	440	356	210	205	225	3461
Station 2	195	195	230	240	290	465	490	465	340	220	190	210	3530
Station 3	205	210	210	226	275	450	500	500	379	242	200	220	3617
Station 4	190	185	220	250	289	500	525	520	400	240	200	215	3734
Mean	192.5	195	220	239	288.5	458.75	497.5	481.25	368.75	228	198.75	217.5	3585.5
Copepoda													
Station 1	170	172	186	185	200	250	256	250	178	145	150	185	2327
Station 2	175	170	180	190	210	260	265	260	195	160	155	190	2410
Station 3	160	185	185	178	190	240	245	240	185	145	140	185	2278
Station 4	165	160	150	175	190	225	225	210	190	132	130	188	2140
Mean	167.5	167.5	175.25	145.6	197.5	243.75	247.75	240	187	145.5	143.75	187	2288.75
Cladocera													
Station 1	155	155	160	140	185	215	205	212	210	145	152	160	2094
Station 2	140	160	152	136	175	195	200	222	210	165	156	156	2067
Station 3	142	160	149	142	162	210	201	210	200	155	145	142	2018
Station 4	130	125	132	140	145	201	204	205	198	120	125	140	1865
Mean	141.75	150	148.25	139.5	166.75	205.25	202.5	212.25	204.5	146.25	144.5	149.5	2011

Contd...

Table 21–*Contd...*

	Nov	*Dec*	*Jan*	*Feb*	*Mar*	*Apr*	*May*	*Jun*	*Jul*	*Aug*	*Sep*	*Oct*	*TOTAL*
Protozoa													
Station 1	105	110	160	125	140	145	125	120	135	82	70	80	1397
Station 2	101	110	132	110	130	125	130	110	120	90	85	75	1318
Station 3	95	100	125	121	125	123	132	100	112	95	88	76	1292
Station 4	85	80	110	102	110	95	110	98	100	90	85	78	1143
Mean	96.5	100	131.75	114.5	126.25	122	124.25	107	116.75	89.25	82	77.25	1287.5

Table 22: Annual Variation in Zooplankton Composition in Tighra Reservoir from November 2010 to October 2011

	Number of Organisms	*Percentage*
Rotifera	3585.5	39.088
Copepoda	2288.75	24.95
Cladocera	2011	21.92
Protozoa	1287.5	14.036

Table 23: Group-wise Seasonal Variation in Zooplankton Composition in Tighra Reservoir from November 2010 to October 2011

Season	*Rotifera*	*Copepoda*	*Cladocera*	*Protozoa*
Summer	1726	929	786.75	479.5
Monsoon	1013	663.25	644.75	365.25
Winter	846.5	655.85	579.5	625.5

Table 24: List of Zooplanktons Recorded in Tighra Reservoir from November 2010 to December 2011

ROTIFERS		
	Brachionus angularis	*B. quadridentatus*
	B. falcatus	*B. calyciflorus*
	B. caudatus	*Keratella tropica*
	Lepadella patella	*Polyarthra* sp.
	Filina longista	*Rotaria* sp.
CLADOCERANS		
	Daphnia pulex	*Moina micrura*
	Macrothrix sp.	*Ceriodaphnia reticulata*
COPEPODS		
	Naupilus sp.	*Mesocyclops hyalinus*
	Cyclopoid sp.	*Mesocyclops leukarti*
PROTOZOA		
	Diffugia muriformis	*Arcella discoides*

Copepods were represented by *Naupilus* sp., *Mesocyclops hyalinus*, *Cyclopoid* sp. and *Mesocyclops leukarti*. Copepod was the second largest group of zooplanktons, representing 24.95 per cent of the total population of zooplanktons. Maximum number of copepods were collected during

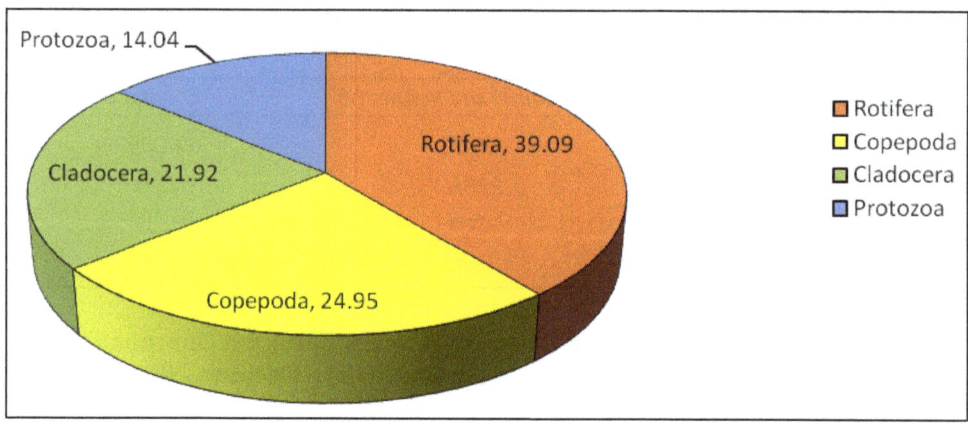

Figure 25: Annual Variation in Zooplankton Composition at Tighra Reservoir, Gwalior.

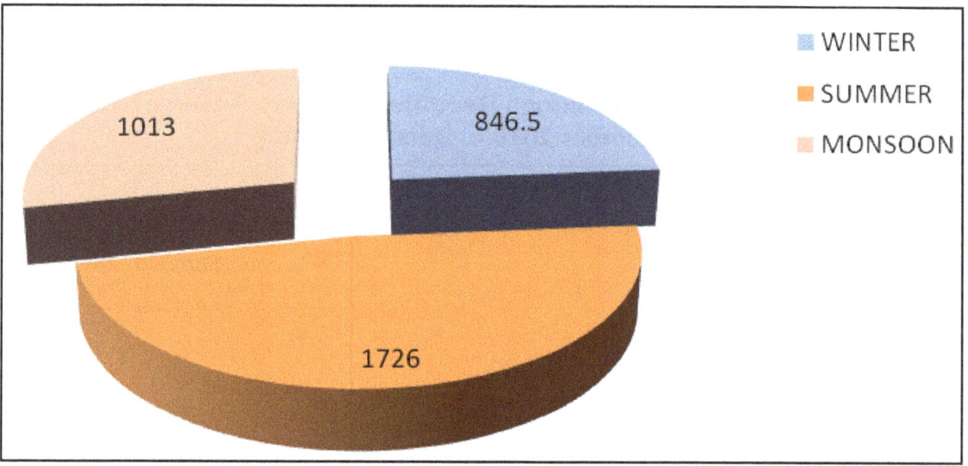

Figure 26: Seasonal Variations in Rotifera (Cells/lit) at Tighra Reservoir, Gwalior for November 2010 to October 2011.

May 2011(265 cells/lit) at Station 2. The number was minimum (130 cells/lit) in September 2011.

Group Cladocera was represented by four genera: *Daphnia pulex, Moina micrura, Macrothrix sp.* and *Ceriodaphnia reticulata.* The numbers was found to be more during the months April, May, June and July 2011. Total number of Cladocera, recorded during the study period was 2011 cells/lit which represented 21.92 per cent of total population of zooplanktons. The zooplankton mass varied from season to season. It was higher during April, May and June (summer), followed by monsoon and lowest during winter.

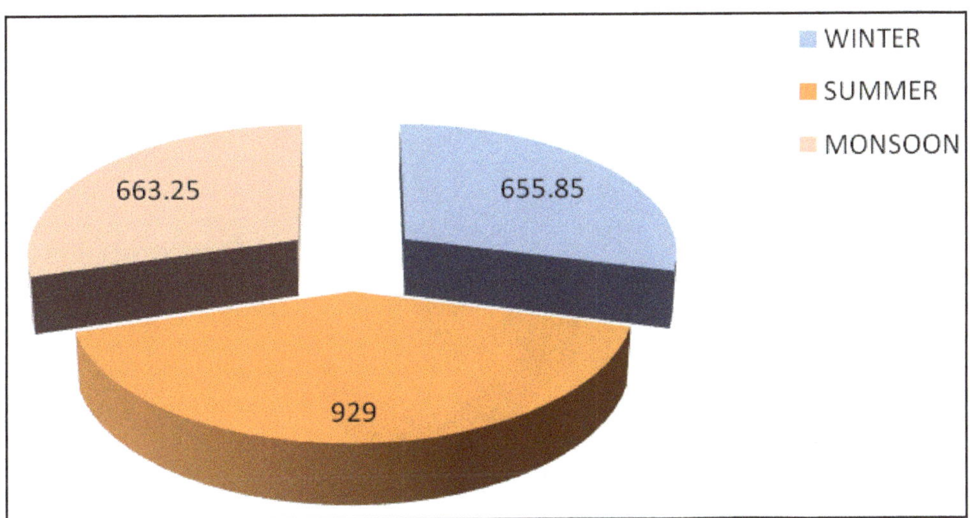

Figure 27: Seasonal Variations in Copepoda (Cells/lit) at Tighra Reservoir, Gwalior for November 2010 to October 2011.

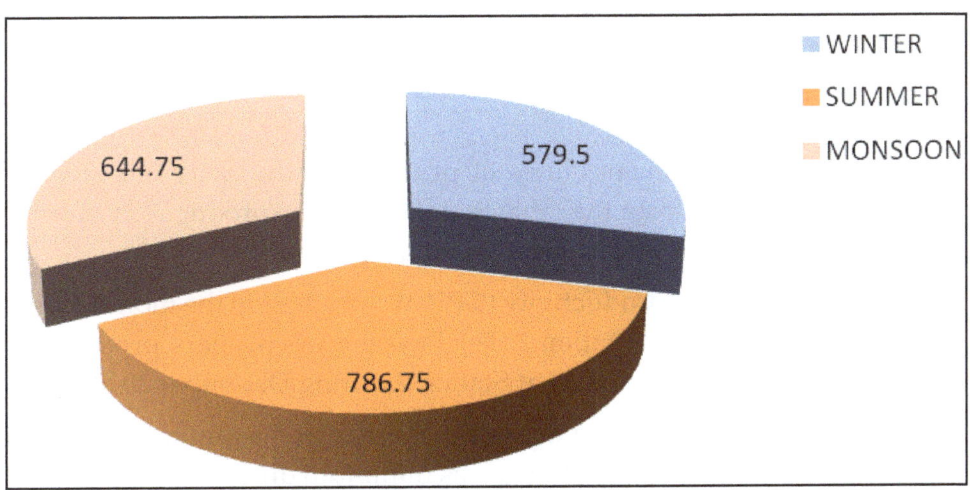

Figure 28: Seasonal Variations in Cladocera (Cells/lit) at Tighra Reservoir, Gwalior for November 2010 to October 2011.

The group Protozoa was mainly represented by *Diffugia muriformis* and *Arcella discoides*.Total protozoa, found during the period of study, was 1287.5 cells/lit. The maximum number was of *Diffugia*. Protozoans were collected in maximum number (160 cells/lit) during January2011 at Station 1 and the minimum number (70 cells/lit) occurred during September 2011 also at the same station. Protozoa represented 14.04 per cent of the total population of zooplanktons recorded in the Tighra reservoir.

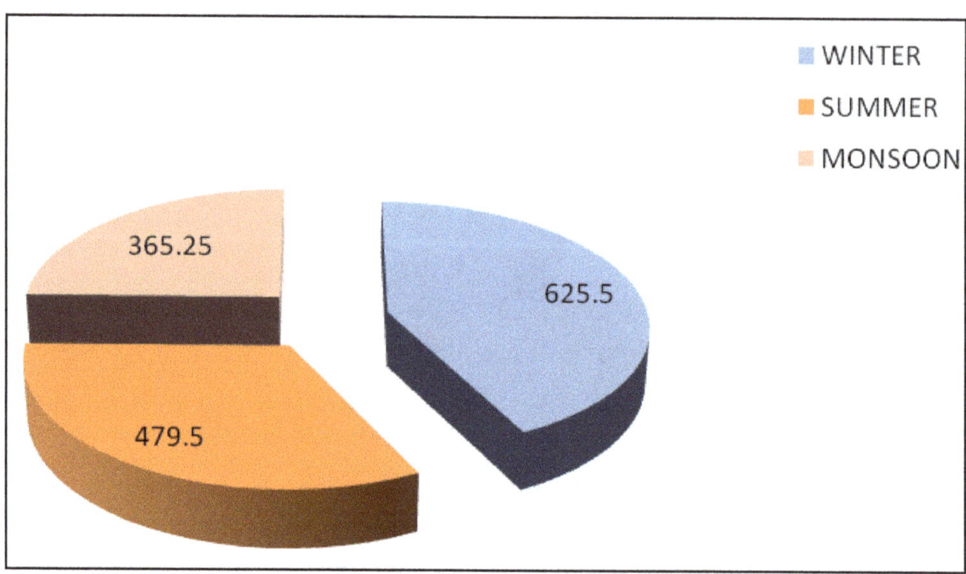

Figure 29: Seasonal Variations in Protozoa (Cells/lit) at Tighra Reservoir, Gwalior for November 2010 to October 2011.

Productivity (Table 25)

Monthly and seasonal values of primary productivity are given in table 23. It is evident from the table that the primary productivity showed a seasonal variation at Tighra reservoir at all stations.

The gross primary productivity (GPP) was highest in the month of June 2011 (0.828gC/m^3/hr) at Station 2. The lowest gross primary productivity (0.154 gC/m^3/hr) was recorded at Station 2 during December 2010.

At Station 1, the gross primary productivity was maximum during June 2011 being 0.776 gC/m^3/hr, while it was minimum during December 2010, being 0.156 gC/m^3/hr. At Station 3, the gross primary productivity was maximum during June 2011 (0.677 gC/m^3/hr) and minimum (0.208 gC/m^3/hr) was recorded during February 2011. Station 4 showed maximum gross primary productivity during June (0.67 gC/m^3/hr) and minimum (0.16 gC/m^3/hr) during December 2010.

Gross primary productivity showed seasonal variations *i.e.* being highest during summer, lower during monsoon and lowest during winter at all four sampling stations.

Table 25: Monthly Variations in Primary Productivity of Tighra Reservoir, Gwalior for November 2010 to October 2011

Month	*Station 1*			*Station 2*			*Station 3*			*Station 4*		
	GPP	*NPP*	*CP*	*GPP*	*NPP*	*CP*	*GPP*	*NPP*	*CP*	*GPP*	*NPP*	*CP*
Nov. 10	0.203	0.109	0.112	0.213	0.067	0.175	0.416	0.208	0.25	0.321	0.104	0.118
Dec. 10	0.156	0.104	0.062	0.154	0.093	0.068	0.404	0.226	0.075	0.16	0.098	0.105
Jan. 11	0.322	0.213	0.143	0.312	0.187	0.15	0.291	0.166	0.102	0.296	0.177	0.143
Feb. 11	0.343	0.161	0.218	0.328	0.161	0.2	0.208	0.102	0.218	0.312	0.156	0.187
Mar. 11	0.375	0.148	0.275	0.368	0.151	0.25	0.223	0.109	0.137	0.364	0.166	0.287
Apr. 11	0.421	0.265	0.187	0.468	0.265	0.181	0.348	0.234	0.135	0.427	0.276	0.181
May 11	0.541	0.328	0.256	0.546	0.338	0.25	0.567	0.412	0.261	0.55	0.338	0.259
Jun. 11	0.776	0.51	0.318	0.828	0.489	0.406	0.677	0.312	0.437	0.67	0.372	0.36
Jul. 11	0.442	0.177	0.318	0.51	0.244	0.318	0.343	0.228	0.163	0.521	0.242	0.341
Aug. 11	0.437	0.072	0.416	0.416	0.135	0.312	0.375	0.288	0.231	0.407	0.1465	0.307
Sep. 11	0.411	0.192	0.262	0.348	0.114	0.281	0.453	0.262	0.218	0.338	0.132	0.241
Oct. 11	0.333	0.11	0.132	0.312	0.255	0.068	0.27	0.187	0.162	0.295	0.127	0.111
Average	0.396667	0.199083	0.224917	0.40025	0.20825	0.221583	0.38125	0.227833	0.199083	0.388417	0.194542	0.22
Summer	0.52825	0.31275	0.259	0.5525	0.31075	0.27175	0.45375	0.26675	0.2425	0.50275	0.288	0.27175
Monsoon	0.40575	0.13775	0.282	0.3965	0.187	0.24475	0.36025	0.24125	0.1935	0.39025	0.161875	0.25
Winter	0.256	0.14675	0.13375	0.25175	0.127	0.14825	0.32975	0.1755	0.16125	0.27225	0.13375	0.13825

Table 26: Correlation Matrix of Physico-chemical and Biological Parameters of Tighra Reservoir from November 2010 to October 2011

	Rainfall	*Temp.*	*pH*	*DO*	*Cl*	*Alk.*	*Hard.*	*CO_2*	*Nitrate*	*Phos.*	*Trans.*	*Turb.*	*Cond.*	*GPP*	*NPP*	*CP*	*Phyto.*	*Zoo.*
Atm. Temp.	0.441	0.855	0.63	–0.87	0.424	0.261	0.857	0.696	–0.11	–0.18	–0.87	0.703	0.491	0.768	0.665	0.707	0.628	0.473
Rainfall		0.632	0.25	–0.403	0.024	–0.33	0.3408	0.758	–0.016	–0.44	–0.691	0.806	0.717	0.467	0.224	0.6731	0.079	–0.076
Temp.			0.49	–0.868	0.520	0.2088	0.8483	0.761	–0.066	–0.0521	–0.765	0.666	0.351	0.880	0.779	0.7587	0.673	0.428
pH				–0.775	0.648	0.5729	0.7582	0.477	–0.423	0.1444	–0.598	0.504	0.334	0.650	0.557	0.6961	0.633	0.594
DO					–0.739	–0.5777	–0.98	–0.722	0.278	–0.1743	0.7827	–0.66	–0.286	–0.936	–0.896	–0.778	–0.87	–0.7
Cl						0.7933	0.733	0.212	–0.151	0.6499	–0.223	0.165	–0.185	0.645	0.726	0.4189	0.92	0.907
Alk.							0.5566	0.122	–0.273	0.7885	–0.074	–0.022	–0.391	0.427	0.585	0.1856	0.774	0.754
Hard.								0.679	–0.355	0.1483	–0.75	0.602	0.255	0.925	0.904	0.7476	0.845	0.679
Free CO_2									–0.249	–0.2402	–0.886	0.807	0.550	0.73	0.608	0.7121	0.393	0.102
Nitrate										0.0391	0.2713	–0.275	–0.163	–0.363	–0.274	–0.533	–0.04	0.013
Phos.											0.3985	–0.403	–0.773	0.100	0.303	–0.172	0.538	0.606
Trans.												–0.916	–0.739	–0.751	–0.582	–0.807	–0.42	–0.24
Turb.													0.829	0.653	0.439	0.8336	0.308	0.206
Cond.														0.257	0.006	0.5825	–0.09	–0.13
GPP															0.936	0.8271	0.771	0.553
NPP																0.5921	0.872	0.659
CP																	0.444	0.311
Phyto.																		0.911

Highest net primary productivity (NPP) was recorded during June 2011 at Stations 1(0.51 gC/m^3/hr). At Station 3 it was maximum (0.412 gC/m^3/hr) in May 2011. Minimum NPP was recorded at Station 2. At Station 2 minimum net primary productivity (0.067 gC/m^3/hr) was recorded during November 2010. Net primary productivity was minimum at Station 1 being (0.072 gC/m^3/hr) during August 2011. Station 3 showed minimum net primary productivity during February 2011. It was 0.102 gC/m^3/hr. At station 4 maximum NPP was recorded (0.372 gC/m^3/hr) during June 2011, while it was minimum (0.098 gC/m^3/hr) during December 2010.Seasonal variations were also observed in net primary productivity being highest during summer,lower during monsoon at Stations 2,3 and 4 and lowest during winter. At Station 1, net primary productivity was recorded to be slightly higher during winter.

The community respiratory consumption varied between 0.062 gC/m^3/hr and 0.416 gC/m^3/hr at Station 1, between 0.068 gC/m^3/hr and 0.406 gC/m^3/hr at Station 2, between 0.075 gC/m^3/hr and 0.437 gC/m^3/hr at Station 3 and between 0.105 gC/m^3/hr and 0.341 gC/m^3/hr at Station 4. Community respiratory consumption, too, was higher during summer than winter.

6

Discussion

Water quality plays an important role in the growth of aquatic animals and their distribution and abundance. The water quality standards, below and above the optimum level, may lead to either stress or death among the aquatic animals. The water quality mainly depends on physical, chemical and biological parameters. Organic and chemical pollutions may also change the quality of water.

Physico-chemical Studies

Physical Parameters

Colour

Colour of the water depends upon many factors such as suspended matter, pH, climate, organic matter and photosynthetic activity. Of these, the materials in solution and particulate matter, both living and nonliving, are important. Some ponds appear to be reddish due to suspended inorganic particles, eroded from surrounding soil. Ganapati (1956) observed reddish colour of water in Hope reservoir throughout his studies which indicated excessive amount of inorganic matter in the water. Verma and Shukla (1968) observed the colour of Devilkin at Deoband to be greenish. Verma (1969) observed the water of Tekanpur reservoir to be transparent, throughout the year except for monsoon months when it became turbid due to inflow of slit into the reservoir along with runoff water. The colour of the water of Tighra reservoir was more or less transparent throughout the study. During monsoon, water was slightly turbid and was about 10 colour units. But the colour was under the standard range. No visible particle or colour was observed. Kaushik and Saksena (1999) reported the water of Motijheel to be transparent during winter, grayish during early monsoon and green during post monsoon and summer season. The grey colour during summer might be due to growth of phytoplanktons, during summer month. Saksena *et al.* (2008) found water of Chambal river, in National Chambal sanctuary, to be transparent to very turbid.

Temperature

Temperature is an important factor in water bodies. It affects chemical and biological reactions in water. No other single factor has so much profound influence on physico-chemical, biological, metabolic and physiological behavior of aquatic ecosystem than temperature (Welch, 1952). It also reflects on the dynamics of living organisms (Chandler, 1942). A rise in water temperature accelerates chemical reactions, reduces the solubility of gases, alters the taste and odour and elevates the metabolic activities of organisms. Among water quality parameters, water temperature is one of the most important parameters, having profound influence on the biotic communities. Air and water temperature depend on geographical location and metrological conditions such as rainfall, humidity, cloud cover, wind velocity *etc.*

In Tighra reservoir, water temperature varied from 18.4 °C to 35.75°C. Minimum temperature was recorded at Station 4 in January 2011. Maximum water temperature, in Tighra reservoir, was recorded during the months of April, May and June. Increase in water temperature, during summer, is due to decrease in water level and exposure of the water to maximum solar radiations. Similar observations were made by Bhagde (2005), Pawar and Pulle (2005), Pawar *et al.* (2007), Kadam *et al.* (2007), Ganjari and Patil (2008) and Jagtap *et al.*(2011).

Kaushik and Saksena (1999) recorded temperature of Ranital between 17 °C and 36 °C. Khaire *et al.* (2011) observed water temperature in the range of 19.2 °C to 27°C. They observed water temperature to be more than the atmospheric temperature.

Fluctuations in water temperature, in Tighra reservoir, at different seasons of the year followed the seasonal variations in atmospheric temperature. A direct relationship between air and water temperature was also found by Welch (1952), Rao (1955), Saha and Pandit (1986), Balkhi *et al.*(1987),Singh *et al.* (1990), Sharma and Jain (2000), Khanna *et al.*(2003) and Sharma (2005).

The water temperature is a determining factor for the distribution of aquatic organisms and the variation in water temperature may be due to different timings of collection and season. Joshi (2006) recorded water temperature between 16 °C and 30 °C in Ekruk reservoir. He recorded maximum temperature in April and minimum in the month of November. Shastri and Bhogaonkar (2006) recorded water temperature in Talwade reservoir in the range of 18 °C to 26 °C. Chandrashekhar (2006) recorded water temperature between 23° and 32 °C in Kondakarla lake in Visakha district, Andhra Pradesh.

Jha and Barel (2003) observed a direct relationship between dissolved oxygen and temperature in their studies on Mirik Lake in Darjelling. Kamal *et al.* (2007) also made a similar observation in their study in Mouri River, Khulna, Bangladesh. Lokhande *et al.* (2010) observed a positive correlation between water temperature and dissolved oxygen in Dhanegaon reservoir in Osmanabad district of Maharasthra. In the present study, a negative relation was observed between water temperature and dissolved oxygen. A negative correlation was observed between water temperature and transparency. Hussainy (1967) and Adoni (1985) also noticed an inverse relationship between temperature and visibility.

Turbidity

Turbidity is an expression of light scattering and light absorbing properties of water. It is caused by the presence of suspended matter such as clay, slit, colloidal organic particles and planktons. Turbidity may also be caused by bottom sediments. It has been greatly recognized as a valuable limiting factor in the biological productivity of a water body. Turbidity interferes with the penetration of light in the water and is a crucial factor, regulating the growth of phytoplanktons.

Turbidity of Tighra reservoir varied from 5.775 NTU to 12.15 NTU. Turbidity was highest during monsoon and minimum during winter. Zafar

(1966) showed a positive correlation between temperature and turbidity. In the present study also, a positive correlation was noticed between temperature and turbidity. It is because of the fact that the particles in water absorb heat and radiate the heat in surrounding water. Chandrashekhar (2006) observed turbidity of Kondakarla Lake to be in the range of 4 to 540 NTU. He recorded exceptionally high turbidity during summer. The high turbidity, during summer, could also be due to excessive growth of phytoplanktons.

Alam *et al.* (2007) observed Surma River to be highly turbid in monsoon. Turbidity was recorded between 1 to 178 NTU by Saksena *et al.* (2008) in Chambal river in M.P. Jagtap *et al.* (2011) recorded turbidity to be between 227 NTU and 274 NTU in Sina-Kologaon reservoir. Maximum turbidity was recorded during summer, while it was minimum during winter. Similar results were reported by Trivedi (1984), Pawar (2006) and Rajshekhar *et al.* (2006).

Transparency

Transparency of water depends on total solids, total dissolved solids and total suspended particles of the water. Transparent waters allow more light penetration which has far reaching effects on all aquatic organisms including their development, distribution and behavior. Transparency is dependent on turbidity which is directly proportional to the amount of suspended matter. Transparency and turbidity play important role in energy dynamics of an aquatic ecosystem. Sreenivasan (1968) reported that waters with low transparency were not suitable for fish growth. Tighra reservoir showed good transparency of water. Transparency of the water varied from 152.75 cm to 211.5 cm. Maximum transparency was recorded during winter and it was minimum during summer and monsoon. Transparency showed an inverse relationship with turbidity. It showed a negative correlation with conductivity. The minimum transparency during rainy season was due to high turbidity of water caused by suspended slit and organic debris. Balkhi *et al.* (1987), Khanna (1993), Sharma (2006, 2007) also reported similar observations. Low transparency during summer may be attributed to the growth of phytoplanktons. Jana (1979) and Bhatt and Negi (1985) also reported similar findings. Sakhare (2005) recorded transparency between 15 and 43 cm. He recorded low values in the month of September which might be due to low and moderate velocity of winter.

Sharma (2005) found maximum transparency of the water of Makroda reservoir, Guna during winter and it was minimum during rainy season. Salgotra *et al.* (2007) observed water transparency between 49.3 and 58.3 cm in Hataikheda reservoir, near Bhopal. They observed maximum transparency during pre-monsoon period. Buktar and Sakhare (2011) observed highest value of transparency, in summer, in the month of May and lowest in winter, in December. Roy *et al.*(2015) observed transparency between 20 cm and 50.10 cm Seep river at Sheopur,Madhya Pradesh.

Conductivity

Conductivity is the numerical expression of ability to carry electric current which in turn depends on the ionic strength. It is an indicator of ionic composition. Conductivity of Tighra reservoir ranged between 272.5 µS/cm and 408.5µS/cm. In the present study, conductivity showed a positive correlation with turbidity. Conductivity was highest during summer which decreased during monsoon and was minimum during winter. Mukherji and Nandi (2006) also recorded conductivity to be more during summer months. Chandrashekhar (2006) recorded water conductivity between 430 and 1720 µs/cm in Kondakarla lake. Khaire *et al.* (2011) reported water conductivity of Mehakari reservoir in Beed District between 252 µmhos/cm and 321 µmhos/cm. They also recorded maximum conductivity during summer and minimum during winter. High conductivity during summer, as recorded in the present study, might be due to increased chlorides and dissolved solids, due to evaporation of water, resulting in increased concentration of salts. Low conductivity, during rainy months may be because of dilution by rain water.

Chemical Parameters

pH

Variation in pH is an important parameter in water bodies since most of the aquatic organisms are adapted to a narrow range of pH and do not withstand abrupt changes. In the present study, the average pH of the water ranged from 6.85 to 7.72 throughout the study period. The maximum pH (8) was recorded during June 2011 at stations 1 and 4. Minimum value (6.6) was recorded at station, 4 during November 2010. Abbasi *et al.*(1996) recorded pH in the range of 6.04 to 7.5 in Kuttadi lake. Buktar and Sakhare (2011) recorded pH value between 8 and 8.4 with a mean of 8, in Wan Reservoir in Maharashtra.

Jagtap *et al.* (2011) recorded pH between 7 and 7.9. They recorded maximum pH value during summer season and minimum during monsoon and rainy season. Seasonally, in Tighra reservoir pH values were higher during summer season. Generally, pH increased in summer and decreased in monsoon and winter, The decrease in pH, during winter, could be due to increase in photosynthesis while the reduction of pH during monsoon could be attributed to the monsoon rains which caused input of freshwater.

Saxena (1992), Pawar and Pulle (2005), Bhagde (2006), Rajlakshmi and Krishanamoorthy (2007) and Jayabhae *et al.* (2008) also supported these findings. In Mehakari reservoir pH varied from 8.1 to 8.7 (Khaire *et al.*, 2011). pH was found to be minimum in the winter and maximum in summer month.

But Tripathi and Pandey (1990) observed maximum pH value during rainy season, in a pond water in Kanpur, Uttar Pradesh. Sanal Kumar *et al.* (2011) observed lowest pH in summer in Achencovil River, Kerala. They attributed the lower values, observed during summer, to a slight increase in CO_2. Chandrashekhar (2006) recorded pH values between 6.9 and 8.9, with maximum values recorded in monsoon and minimum values in summer.

Sakhare (2006) recorded pH between 7.4 and 8.5 in the water of Jawalgaon reservoir in Solapur, Maharastra. But he did not observe any definite seasonal variation. Singhai (1986) obtained a direct relationship between water temperature and pH. Roy *et al.* (2015) recorded pH between 7.45 and 8.80 in their studies on Seep river. Vardhan *et al.* (2015) observed pH of Tatapani Spring of Jammu varying from 8.2-9.2 which was greater than the required desirable limit (6.5-8.5) recommended by Indian standard specifications for drinking water (IS).

The desirable pH of drinking water is 7 to 8.5 (WHO, 1984). The results show that the water of Tighra reservoir is suitable for drinking purpose and is under the standard range.

The pH between 6 and 9 is most suitable for fish culture and pH more than 9 is unsuitable (Ellis, 1937 and Swingle, 1967). Verma and Shukla (1968) observed that alkaline waters support large amount of biota. According to Jhingran and Sugunam (1990) reservoirs with pH range between 6 and

8.5 are productive in nature, reservoirs with pH more than 8.5 are highly productive and reservoirs with pH less than 6 are less productive reservoirs. pH showed a negative correlation with dissolved oxygen and a positive correlation with alkalinity, turbidity, chloride and hardness.

Dissolved Oxygen

Many gases are dissolved in natural waters. Dissolved oxygen is one of the most important parameters in water quality and is an index of physical and biological processes going on in the water. It regulates metabolic processes of the organisms and also of the community as a whole. It acts an indicator of trophic status and magnitude of eutrophication. Dissolved oxygen, along with turbidity, gives information about the nature of an aquatic ecosystem better than any other chemical parameter (Hutchinson, 1957). There are two main sources of dissolved oxygen in water- diffusion from air and photosynthetic activities within the water. Diffusion of oxygen from atmosphere is a physical phenomenon and depends on solubility of oxygen, which is often affected by factors like temperature, water movements and salinity.

In Tighra reservoir, average dissolved oxygen varied from 5.425 to 8.125 mg/lit. The maximum value (8.2 mg/lit) was recorded at station 3 during November 2010 and at Stations 1 and 4, during December 2010. The lowest value was recorded at Station 2 during June 2011. Seasonally, high values were recorded during winter season at all four stations. The values decreased gradually from February till June. Minimum values were recorded during May- June. Munawar (1970) also observed higher values of dissolved oxygen in the month of April and during winter and low values during summer. Raghuwansi (2005) recorded dissolved oxygen in the range of 12.8 to 18.6 mg/lit at lower lake Bhopal. He recorded minimum dissolved oxygen in April and maximum in June. Salogtra *et al.* (2005) observed significant seasonal variation in dissolved oxygen of Hataikheda reservoir, near Bhopal, Madhya Pradesh. Somal *et al.* (2005) also recorded maximum dissolved oxygen during the winter season and minimum during monsoon in Hirakund reservoir. Adebies (1981), Sakhare and Joshi (2003), Pawar and Pulle (2005), Kadam *et al.*(2007), Srinivasarao *et al.* (2007), Ganjari and Patil (2008). Kamble and Mulle (2008), Ayoade (2009) and Lokhande *et al.* (2010) also reported similar results. Saksena *et al.*(2008) recorded dissolved oxygen between 4.86 to 14.59 mg/lit in Chambal river in Madhya Pradesh.

Buktar and Sakhare (2011) recorded dissolved oxygen in the range of 1.32 mg/lit to 6.2 mg/lit. The maximum value was recorded in the month of December 2010 and lowest in May 2011. Jagtap *et al.* (2011) recorded dissolved oxygen in the range of 7.3 to 8.6 mg/lit with maximum values in winter season, moderate in monsoon and minimum in summer. Sanal Kumar *et al.* (2011) also observed similar results. Khaire *et al.* (2011) recorded dissolved oxygen between 3.8 and 7 mg/lit. They recorded peak values during winter and least in summer.

The high value of dissolved oxygen, observed during winter might be due to increase in photosynthetic activity which liberates a considerable amount of oxygen in water. At low temperature the solubility is high and there is less degradation of organic substance during winter. Low dissolved oxygen levels, during summer, indicates that the oxygen replenishment rate is lower than utilization. The decomposition of organic matter and microbial activity is higher in warm weather which causes depletion of oxygen during summer (Morissote *et al.,* 1978).

But Chandrashekhar (2006) reported comparatively higher values of dissolved oxygen during monsoon period which he attributed to the stable abiotic conditions and higher biomass, stimulating the rate of photosynthesis. Shastri (2005) observed a positive correlation between temperature and dissolved oxygen in a small percolation tank. Agarwal and Thaliyal (2005) found an inverse relationship between dissolved oxygen and temperature. A negative correlation was observed between temperature and dissolved oxygen of water of Tighra reservoir. Kamal *et al.* (2007) also observed a direct relationship between dissolved oxygen and water temperature.

Free CO_2

Respiratory activities of aquatic organisms and process of decomposition are important sources of CO_2 in water bodies. Free CO_2 is added to aquatic ecosystem by directly being mixed from atmosphere. The presence or absence of free carbon dioxide in water is mostly governed by its utilization by algae during photosynthesis and also through its diffusion from air. Free carbon dioxide serves to buffer the environment against rapid shifts in the acidity or alkalinity and also regulates biological processes in aquatic communities. Thus it is an important factor which contributes to the quality of natural water. Sakhare (2005) measured free CO_2 between 0 and 4.1 mg/lit, with an average of 2.8 mg/lit, in Hingini (Pangaon) reservoir. The maximum concentration was recorded during August and free CO_2 was found to be totally absent during November.

Free CO_2 was recorded in the range of 4.15 mg/lit to 7.57 mg/lit in Tighra reservoir. Free CO_2 was found more during monsoon period and minimum during winter. Free CO_2 showed a negative correlation with transparency and a positive correlation with turbidity and conductivity.

Saksena *et al.* (2008) recorded free CO_2 between 0 to 16.5 mg/lit in Chambal river. Buktar and Sakhare (2011) measured free CO_2 in the range of 0 to 4.3 mg/lit. Jagtap *et al.* (2011) recorded free CO_2 between 0 and 0.3 mg/lit. The maximum value was recorded during summer while nil value was recorded during winter season. Kumar (1995), Kadam *et al.* (2007) and Ganjari and Patil (2008) reported similar results. Sanal Kumar *et al.* (2011) recorded high free CO_2 during summer. Increased rate of organic compounds decomposition in summer causes the liberation of CO_2 in water and consequently elevated the free CO_2 level in water in summer months. Sharma (2006) recorded high free CO_2 during monsoon and low or absence of CO_2 during summer. Singhai *et al.* (1990), Badola and Singh (1981) reported similar findings. Chandrashekhar (2006) reported high values of free CO_2 during summer. High values during summer may be due to increased decomposition of dead organic matter with a rise in temperature.

Total Hardness

Hardness is frequently used as an assessment of water quality. Water hardness is generally governed by calcium and magnesium contents of water.

In the present study, hardness was recorded between 66.25mg/lit and 137 mg/lit. Saksena *et al.* (2008) recorded total hardness between 42mg/lit and 140 mg/lit of Chambal river. Elamci *et al.*(2008) measured hardness of lake Uluabat to be between 14.61 mg/lit and 140.94 mg/lit. Shastri (2005) recorded total hardness between 130mg/lit and 290 mg/lit in a small percolation tank.

WHO's (1984) permissible limit for total hardness of water for drinking purpose is 150 mg/lit and ISI (1983) limit is 300 mg/lit. Total hardness, in the range of 150-300 mg/lit, indicates the water to be very hard (Todd,1995). The total hardness of Tighra reservoir was between 66.25mg/lit and 137 mg/lit which was under permissible limits. This indicates that the water of

Tighra reservoir is suitable for drinking purpose. Buktar and Sakhare (2011) found total hardness to be maximum (59 mg/lit) in December and minimum (40 mg/lit) during May. But in the present investigation total hardness was recorded to maximum during summer and minimum during winter. Similar observations were reported by Joshi (2006), Sakhare (2006) and Khaire *et al.* (2011). Joshi (2006) reported total hardness of Ekruk reservoir between 59 and 104 mg/lit, indicating the suitability of water for drinking purpose. The hardness was minimum 59 mg/lit in December and maximum (104mg/lit) in March. Kadam *et al.* (2006) recorded total hardness in the range of 100 to 109 mg/lit, which was well below the standard range of hard water. Sakhare (2006) recorded maximum hardness in May and minimum during January. Khaire *et al.* (2011) reported total hardness of Mehakari reservoir between 76 and 130 mg/lit. The values were maximum in summer, followed by monsoon and least in winter.Vardhan *et al.* (2015) reported total hardness ranging from 440 ppm to 475 ppm in Tatapani Spring of District Rajouri – Jammu. Optimum hardness for fish culture has been reported to be around 150 mg/lit (Das1996). Accordingly, the water of Tighra is suitable for fish culture.

Alkalinity

Alkalinity is the capacity to neutralize a strong acid. It is a measure of the ability of the water to absorb H^+ without significant change in pH. In natural waters, most of the alkalinity is caused due to CO_2. The free CO_2 dissolves in water to form carbonic acid in large quantities. Alkalinity imparts bitter taste to the water. The average alkalinity ranged between 53.76 mg/lit and 145.5 mg/lit in Tighra reservoir. Salgotra *et al.* (2005) observed total alkalinity in the range of 24.5 to 65.7 mg/lit in Hataikheda reservoir. Sakhare (2005) measured total alkalinity between 49 and 69.8 mg/lit with an average value of 52.5 mg/lit. The maximum value was recorded during November while it was minimum during March. In Jawalgaon reservoir, Sakhare (2006) observed total alkalinity to be more than 100 mg/lit throughout his study. According to Jhingran (1988), low alkalinity is not conducive for good productivity since highly productive waters have alkalinity more than 100 mg/lit. Buktar and Sakhare (2011) recorded total alkalinity in the range of 243 mg/lit to 310 mg/lit in their investigation on Wan reservoir. Marked seasonal variations were shown by Jagtap *et al.* (2011), in total alkalinity. They reported total alkalinity between 102 and 145 mg/lit. Maximum values were recorded during winter, moderate during summer and minimum during monsoon. Similar observations were made by Shastry *et al.* (1999), Sakhare and Joshi (2003), Kadam *et al.* (2007), Mane and Pawar (2007) and Aher *et al.*(2007). Agarwal and Thapliyal (2006) also obtained maximum alkalinity during winter in Bhilangana.

In Tighra reservoir maximum alkalinity was recorded during summer, followed by winter and minimum during monsoon season. Sharma (2005, 2006) reported highest values of alkalinity during summer season. Increase in temperature and decrease in water level can produce rise in alkalinity during summer (Varghese *et al.*, 1992). Khaire *et al.* (2011) observed total alkalinity in the range of 115 mg/lit to 185 mg/lit. They observed maximum total alkalinity during summer and minimum in winter.

Chloride

Chloride is an indicator of contamination of water with animal and human wastes. In water bodies chloride content is affected by influx of drainage and evaporation of water during high temperature. In the present study, chloride content ranged between 11.3 mg/lit and 40.7 mg/lit. Low chloride contents indicate that the water is not polluted and is safe for drinking purpose. The chloride content was high during summer and low during winter. Sharma (2005, 2006) also reported high chloride content during summer and low during winter. Similar results were also reported by Mukherji and Nandi (2006), Sakhare (2011), Jagtap *et al.*(2011) and Khaire *et al.* (2011).

Jagtap *et al.* (2011) found highest chloride value during summer, moderate in winter and lowest during monsoon season in their studies on Sina-Kolegaon reservoir. Buktar and Sakhare (2011), too, found higher values of chloride in the month of May. In Mehakari reservoir chloride concentration varied from 17.04 to 29.82 mg/lit (Khaire *et al.*, 2011). Joshi (2006) observed maximum chloride content (49.2mg/lit) in April and minimum 19.3mg/ lit) in August.

Chandrashekhar (2006) recorded maximum chloride content during monsoon season. Sakhare (2006) recorded higher values of chloride in winter and lower in monsoon. Subamma and Rana Sarma (1992) have discussed the fluctuations in chloride content of various water bodies and have reported minimum concentration in monsoon season.

A positive correlation was observed between chloride, hardness and alkalinity.

Nitrates

In the present investigation nitrate content of Tighra reservoir was recorded between 0.29 mg/lit and 0.81mg/lit. Shastri (2005) recorded nitrate concentration between 0 – 34 mg/lit in his study on a small percolation tank. The low value indicates oligotrophic nature of the tank. Saksena *et al.* (2008) reported nitrate content between 0.008-0.025 mg/lit in Chambal river. In Mehakari reservoir, Khaire *et al.* (2011) observed nitrates in the range of 0.08 mg/lit to 0.52 mg/lit. Sharma (2006) reported concentration of nitrate to be maximum during rainy season in Rampur reservoir, Guna, M.P. Nitrate contents decreased during winter and again increased during summer. The high values of nitrates, during rainy season, may be due to addition of nitrates into the water by runoff water. In Tighra reservoir nitrates concentration was slightly higher during monsoon and winter season than summer months. Chandrashekhar (2006), too, noticed low values of nitrates in summer and high values in monsoon.

The high concentration of nitrate-nitrogen has been reported to be an index of eutrophication (Strom 1930 and Wetzel, 1975). According to Ganapti (1960) tropical waters, particularly unpolluted ones, are deficient in nitrates and concentration beyond 0.15 mg/lit of nitrates is an indicative of eutrophication. Shastri and Bhogaonkar (2006) recorded nitrates in the range of 0-0.19 mg/lit.

Phosphate

According to Schlinder (1971), phosphate is amongst the primary limiting nutrients in ponds and lakes. It is usually present in low concentration in natural unpolluted water. Phosphate content, in Tighra reservoir, was recorded between 0.375 mg/lit and 1.575 mg/lit. Saksena *et al.* (2008) found phosphate content in Chambal river in the range of 0.004 to 0.05 mg/lit. Elmaci *et al.* (2008) recorded total phosphate in lake Uluabat, Turkey between 1.11 mg/lit and 3.01 mg/lit. Shastri (2005) observed low phosphate values (between 0 and 0.28 mg/lit) in a small percolation tank. Khare (2005) recorded phosphate values between 0.10 and 0.26 ppm in Jagat Sagar Pond, Chhatarapur. Khaire *et al.* (2011) recorded phosphate in the range of 0.09 to 0.62 mg/lit in Mahakari reservoir. Shastri and Bhogaonkara (2006) reported phosphate content in the range of 0 to 0.52 mg/lit. The concentration was 0 in April, indicating low productivity of the reservoir. Chandrashekhar (2006) found nil or very low amount of phosphate in Kondakala lake. Roy *et al.*(2015) reported phosphate varying from 6.40 mg/l to 18.30 mg/lit with an average of 9.49 mg/lit in Seep river, Sheopur.

Mukherji and Nandi (2006) reported phosphate content to be higher during pre-monsoon and monsoon. In the present investigation, phosphate content was higher during summer and low during monsoon. High concentration during summer could be due to higher decomposition of organic matter in summer (Mehra,1986 and Kumar, 1995).

Biological Parameters

Phytoplanktons

In aquatic ecosystems, phytoplanktons are the important components which consist of primary producers with photosynthetic activity that constitute the first step in the carbon cycle of water body. The wide distribution of phytoplanktons and their frequent abundance account for the great importance of the phytoplanktons in the food web of aquatic ecosystems. There is a large variation in the composition of planktons in different regions and at different depths and also at different seasons. In water bodies phytoplanktons are represented by algae population. Often one or small groups of species may be dominate, depending upon the nutritional status of the water body.

The most common phytoplankton groups found in freshwater bodies are - Bacillariophyceae, Chlorophyceae, Myxophyceae and Euglenophyceae.

Bacillariophyceae is the most important group of algae. Even though most species are sessile and are associated with littorial substrates. Both unicellular and colonial forms are common among bacillariophyceae. The group is commonly divided into centric diatoms which exhibit radial symmetry, and pinnate diatoms which have bilateral symmetry.

Myxophyceae occur in unicellular, filamentous and colonial forms and most of them are enclosed in mucilaginous sheaths either individually or in colonies. Most of these planktons consist of members of coccoid family (*e.g. Microcystis*), filamentous families (*e.g. Planktothrix*), Nostocaceae (*e.g. Anabaena*) and Rivulariaceae (*e.g. Gleotrichia*). Myxophyceae have circadian rhythm and the capacity for photosynthesis and nitrogen fixation is regulated by a biological clock.

Chlorophyceae form an extremely large and morphologically diverse group of algae that is almost freshwater in distribution. Most green algae belong to order Volvocale (*e.g. Volvox*) and Chlorococales (*e.g. Scendesmus*). Many members are flagellated while others are filamentous *e.g. Spirogyra.*

Although Euglenophyceae are a relatively large and diverse group, few species are truly planktonic. Almost all euglenophyceae are unicellular, lack a distinct cell wall and possess two or three flagella which arise from invagination in the cell membrane. Although some euglenophyceae are unpigmented, most of them are photosynthetic. They are found most often in the shallow water, rich in organic matter such as organically polluted lakes (Yousoff and Patimah, 1994) and farm ponds (Wetzel, 1975).

Pandey *et al.* (1995) recorded four algal groups: chlorophyceae, bacillariophyceae, myxophyceae and euglenophyceae in river Saura with Chlorophyceae being the most dominant group. Harsha and Malammanavar (2004) recorded maximum density of cyanophyceae in their studies on Gopalaswamy pond at Chitradurga, Karnataka. They observed considerable fluctuations in phytoplankton density in relation to environmental variables.

Out of the six major groups of phytoplanktons, recorded from Dal Lake, Kashmir, Jeelani *et al.* (2005) found bacillariophyceae as the most dominant group. They also observed seasonal variations in the distribution of phytoplanktons with bacillariophyceae peak during autumn and chlorophyceae, cyanophyceae, euglenophyceae, chrysophyceae showed peak in summer.

In their study, Kumar and Bohra (2005) found maximum number of phytoplanktons during summer and minimum number in rainy season in the Munshi pond, Jharkhand.

Sakhare (2006) identified 14 species of chlorophyceae, 9 species of bacillariophyceae and 7 species of myxophyceae from Jawalgaon reservoir.

Tiwari and Chauhan (2006) observed chlorophyceae as the most dominant during winter, followed by bacillariophyceae in Kitham Lake, Agra. During summer, cyanophyceae and euglenophyceae were the most dominant groups.

In Tighra reservoir all four groups of phytoplanktons, bacillariophyceae, chlorophyceae, myxophyceae and euglenophyceae were recorded throughout the study period. Bacillariophyceae was the most dominant of all the groups of phytoplanktons. Seasonally, maximum numbers of phytoplanktons were observed during summer and lowest number in rainy season. This variation in phytoplanktons number may be due to high temperature. A positive correlation was also observed between temperature and phytoplanktons in Tighra reservoir.

Several researchers have proposed temperature as a vital factor, responsible for the growth of algae (Ramkrishnaiah and Sarkar, 1982; Verma and Datta Munshi, 1987; Kaushik *et al.*, 1991; Bohra and Kumar, 1999). Wisharad and Mehrotra (1988) reported that proliferation of phytoplanktons from winter to summer could be attributed to progressively increasing water temperature and photoperiod. According to Cabecadas and Brogueira (1987), the growth and photosynthesis of algae are influenced by the pH and alkalinity of water. In the present investigation, on Tighra reservoir, phytoplanktons showed positive correlation with pH, chloride, alkalinity, hardness and phosphate. Among the various groups of phytoplanktons, positive relationship was observed between myxophyceae and between myxophyceae, bacillariophyceae and hardness.

Senthi Kumar and Siva Kumar (2008) identified total 160 species of phytoplanktons in Veeranam lake in Tamil Nadu with bacillariophyceae being the dominant group. The phytoplankton density was high during summer season and low during the winter season.Their studies corroborate with that in present investigation of Tighra reservoir.

Lashkar and Gupta (2009) found a total 34 phytoplanktons, belonging to cyanophyceae, chlorophyceae, bacillariophyceae and euglenophyceae in Chatla flood plain lake, Barak Valley Assam. They recorded highest number of species (29) in pre-monsoon and lowest (23) in winter. In their studies, they observed chlorophyceae to be the most abundant in pre-monsoon and monsoon. Bacillariophyceae and cyanophyceae did not show much seasonal variations.

Synudeen Sahib (2011) recorded 35 species of phytoplanktons in the Parappar reservoir, Kerala. The chlorophyceae represented the maximum density. He observed maximum density of planktons in winter and the minimum during rainy season. Low temperature and velocity, coupled with good transparency of water, may be considered as the factors that favored the optimum growth of phytoplanktons of the Parappar reservoir during winter. Phytoplankton in Mahanadi of Odisha were investigated seasonally by Panigrahi and Patra (2013). Their studies revealed that physico-chemical parameters such as water temperature, pH, total nitrogen, total alkalinity, chloride and phosphate were significantly related to phytoplankton abundance, but total abundance and community structure of phytoplankton were not influenced by environmental factors.

Zooplanktons

All groups of zooplanktons – Rotifera, Cladocera, Copepoda and Protozoa were recorded in the water of Tighra reservoir during the study period. Rotifera was the dominant group throughout the study period. Distribution of zooplanktons also revealed seasonal variations. Rotifera, Cladocera and Copepoda were found in maximum number during summer, followed by winter and minimum during monsoon. In case of protozoa, maximum number was recorded during monsoon, followed by summer and least in winter. Zooplanktons had a positive correlation with phytoplanktons. Positive correlation was also observed among zooplanktons and alkalinity and hardness.

Protozoa is a group of unicellular ciliated or flagellated organisms that feed on either picoplanktons or nanoflagellates and small nanophytoplanktons according to their size.

Rotifers are the most important soft bodied invertebrates having a very short life cycle among planktons. They are influenced by temperature, food and photoperiod.

Cladocerans form a crucial group among zooplanktons which is the most useful nutritive group of crustaceans. The greater significance of Cladocera in the aquatic food chain, as a food for both young and adult fish, was emphasized much earlier (Pennak, 1978). Cladocerans feed on small zooplanktons, bacterioplanktons and algae. They are highly responsive to pollutants and even react against low concentration of contaminants.

Among all the zooplanktons, copepods have the toughest exoskeleton and the longest and strongest appendages which help them to swim faster than any other zooplankton. Ostracodes are mainly bottom dwellers of lake and live on detritus and dead phytoplanktons.

Jhingran (1989) recorded cladoceran population to be most abundant in February, followed by July and October in Ramgarh reservoir in Rajasthan. Sharma and Diwan (1993) studied plankton dynamics of Yeshwant Sagar reservoir in which the Cladocera showed maximum density in June. They reported rotifers to form a dominant group during summer in Yeshwant sagar reservoir.

Deshmukh (2001) recorded 28 species of rotifers from Chhatri Lake of Amravati with maximum number of species in summer. Akin-Oriola (2003) observed rotifera as the most dominant zooplankton in Ogunpa and Ona rivers, Nigeria. The dominance of rotifer was attributed to their short development rate and fish predation on large zooplanktons. Khare (2005) observed an increasing trend in the months of winter season with peak during summer months- March to June. He recorded minimum population during rainy season. Mishra (2005) identified 32 species of zooplanktons, belonging to 16 families and 24 genera from the Morar River, Gwalior. Kadam *et al.* (2006) observed maximum number of rotifer during summer season.

Rajshekhar *et al.* (2010) recorded 24 species of zooplanktons in a freshwater reservoir of Gulberbarga district, Karnataka. Rotifer was the dominant group throughout their study period. Highest count was recorded during summer.

VasanthKumar *et al.* (2011) studied species richness, diversity and evenness of zooplanktons. They recorded 61 species of zooplanktons, in three ponds of Karwar district, Karnataka, with rotifer being the most dominant group. In their study, they observed a positive significant correlation between zooplankton population and physico-chemical parameters like temperature, alkalinity, phosphate, hardness and negative correlation with rainfall and salinity. Quadri (2015) investigated 20 species of zooplankton belonging to different groups *i.e.* Rotifera, Copepoda, Cladocera, Ostrocoda in Kham river, Aurangabad.

Productivity

Primary productivity studies, in an aquatic environment, are not only important for the comparison of productive capacity of the waters belonging to different geographical regions but also for their value in fish culture programmes.

Primary productivity of aquatic environment primarily depends on the photosynthetic activity of the autotrophic organisms which are capable of converting carbon dioxide into organic matter. It can be studied in two forms as Gross Primary Productivity (GPP) and Net Primary Productivity (NPP).

GPP refers to the total organic substance produced by photosynthesis in a definite time. Net primary productivity is the rate of storage of organic matter, in excess of respiratory utilization by phytoplanktons.

In the present investigation gross primary productivity (GPP) was recorded to in the range of 0.828 $gC/m^3/hr$ to 0.154 $gC/m^3/hr$. GPP was maximum during summer while it was minimum during winter (post-monsoon period). A seasonal variation was recorded in the primary productivity of the reservoir. It is generally realized that clear days, in summer months, produce relatively greater productivity than that of cloudy days. Productivity was found to increase from late winter, reaching peak in the late summer or early rains and to declining thereafter. The high rate of productivity, observed during summer, was probably due to bright sunshine, high temperature, low transparency, high phytoplankton density and blooms, as reported by Sreenivasan (1964, 1969), Khan and Siddiqui (1971),Singh and Saha,(1987), Takamura *et al.*(1989) and Bohra and Kumar (2002). The declining tendency of productivity, during rainy season, could be because of the reduction in the size of the productivity zone due to rain, increase in the water level, dilution of nutrient concentration and cloudy weather. Low productivity, observed during winter, could be attributed to low intensity of light, short photoperiod, low concentration of nutrients and less number of phytoplanktons (Khan *et al.*1988; Rai and Sharma 1991; Pati and Sahu,1993; Vijay Kumar,1994; Kumar, 1996,1999; Mishra and Nayak,2000 and Bohra and Kumar,2002).

The minimum GPP, recorded during post monsoon, might be due to low phytoplankton population together with factors like dilution of nutrients, cloudy weather and comparatively low temperature. The highest GPP, recorded during summer was due to increase in temperature and alkalinity. Increase in GPP (after monsoon) could be contributed to the nutrients and organic matter that get accumulated and are available for production.

Khan and Siddiqui (1971) recorded high gross primary productivity between March and May and low values for the rest of the period from a tropical pond in Aligarh. Sukumar (2002) observed nil GPP in summer in Lalbagh tank of Banglore while Eshwarlal and Angadi (2002) recorded absence of GPP duing winter in Jagtap tank in Gulbarg. Maximum GPP, in summer, was also recorded by Jayachandran and Krishanan (2001) and Raj Kumar (2004), while Vijay Kumar *et al.*(2000) and Das (2002) reported maximum GPP during winter. Raj Kumar (2004) recorded complete absence of GPP during monsoon in KAK pond, Coimbatore.

Kumar and Bohra (2005) observed primary productivity in the range of 0.48 to 6.21 gc/m^3/d in Munshi pond and 0.48 to 3.35gc/m^3/d in Raja Dighi pond. The minimum and maximum values of phytoplanktons were observed during winter and summer seasons, respectively. Fasihuddin and Puttaiah (2005) recorded GPP in the range of 0.18 to 43.42 mgCm^{3}d^{-1} in Manikanava reservoir in Karnataka.

Shiddamallayya and Pratima (2007) also recorded maximum productivity (606 gCm^{-3}hr^{-1}) during summer in Papnassh pond of Bidar district which corroborates with the present findings on Tighra reservoir. The maximum GPP was noted during summer season and minimum in rainy season by by Moharana and Patra (2013) in their studies of Bay of Bengal at Digha.

In the present investigation, NPP was recorded in the range of 0.51 gC/m^3/hr to 0.067 gC/m^3/hr. The values were high during summer and low during winter. The higher values, observed during summer, might be due to availability of nutrients and favorable conditions for the setting of algal bloom. Similar findings were reported by Eshwaralal and Angadi (2002) in Jagtap tank and Mali and Gajara (2004) in a fish culture tank in Gujrat. NPP values were found to fluctuate with the values of GPP. This indicates that there was no interference by primary consumers like zooplanktons and others.

Fasihuddin and Puttaiah (2005) recorded NPP between 0.10 and 22.69 mgCm^{3}d^{-1}. Shiddamallayya and Pratima (2007) recorded NPP between -455 gCm^{-3}hr^{-1} to 354 gCm^{-3}hr^{-1} in Bhalki tank in Bidar. Moharana and Patra (2013) observed NPP value varying from 0.015 + 0.002 gmh to 0.845 + 0.072 gmh and showed an increasing trend from the month of September 2011 to February, 2012.

In Tighra reservoir, the community respiratory consumption varied between 0.062 gC/m^3/hr and 0.416 gC/m^3/hr. Community respirations tend to be increasing during summer and decreasing in the monsoon and winter. Vijay Kumar (1995) also made similar observations. But Singh (1990) and Sukumar (2002) reported low CR values during summer in North and South lakes of Jamalpur and Lalbag tank, Banglore, respectively.

Fasihuddin and Puttaiah (2005) recorded CR between 0.20 and 21.2 $mgCm^3d^{-1}$ in Manikanava reservoir. Shiddamallayya and Pratima (2007) recorded variation in Community respiration (CR) in range of 24 $gCm^{-3}hr^{-1}$ to 348 $gCm^{-3}hr^{-1}$ in Papnosh pond. They recorded minimum CR in the month of March and maximum in April in the following year.

High values of CR during summer, as recorded in the present study and as also reported by Shiddamallayya and Pratima (2007), might be because of higher phytoplankton and zooplankton populations.

Interaction between Biotic and Abiotic Factors

The aquatic life in a water body is governed by physico-chemical conditions and biological conditions of the water body. Davis (1955) pointed out that various physico-chemical and biological circumstances must be simultaneously taken into consideration for understanding fluctuations of plankton population.

Interaction between physico-chemical factors and biological factors is observed in the water bodies. Investigations have been made to correlate plankton distribution with physico-chemical parameters. Correlation between physico-chemical factors and planktons has been studied by many workers(Chakarabarty *et al.*, 1957, Arora 1966, Nayak *et al.*1982,Kaushik *et al.*, 1991c, Adholia and.Vyas 1992, Kumar 1995, Joshi *et al.*1996, Akin-oriola 2003, Harsha and Malammanavan,2004; Ayoade *et al.*2009; Lashkar and Gupta 2009;Rajshekhar *et al.*, 2010, Basu *et al.*, 2010). The study of correlation between biotic and abiotic factors is useful in gaining basic knowledge of the trophic status of a water body. The distribution of certain species was correlated with temperature, pH, dissolved oxygen, hardness, alkalinity and inorganic nutrients.

In addition to C, H and O_2, phytoplanktons require some other essential nutrients to grow well. It is generally observed that if concentrations of nitrogen and phosphorous are low, they limit the phytoplankton productivity. Agarwal *et al.* (1990) developed a relationship between nutrients and algal growth. Pandy *et al.* (1995) showed a positive correlation between pH, dissolved oxygen, bicarbonate, phosphate and transparency.

A positive correlation was observed among pH, dissolved oxygen and transparency and Chlorophyceae.

Bhat and Pandit (2005) found a close relationship between physico-chemical characters of water and growth and abundance of phytoplanktons. They observed high growth of phytoplanktons during summer and very low growth during winter. Kumar and Bohra (2005) showed a significant positive correlation between phytoplanktons and pH in Raja Dighi pond, Jharkand.

Hulyal and Kaliwal (2008) showed a significant relation between biotic and abiotic factors. Lashkar and Gupta (2009) observed a highly significant positive correlation between phytoplankton density and transparency ($p<0.01$) and significant positive correlation with total hardness.

Hulyal and Kaliwal (2009) revealed a positive relationship between cyanophyceae with dissolved oxygen, nitrate, phosphate and negative correlation with pH, chloride, rainfall and humidity. Similarly, an inverse correlation was found between bacillariophyceae and rainfall, humidity and phosphate. In the present investigation, phytoplanktons had positive relationship with alkalinity, hardness and phosphate.

Agarwal and Rejwar (2010) found pH, conductivity, hardness and DO to be higher during summer months. Nitrate increased in summer and early monsoon due to higher phytoplankton production.

Synudeen Sahib (2011) observed a close relationship between turbidity and velocity and plankton biomass. A rise in turbidity, during summer and rainfall, leads to silting, disturbances of normal O_2 and CO_2 exchange, consequently an inhibition of photosynthesis of the phytoplanktons. During winter, DO reaches at peak and free CO_2 remains less while a reverse situation occurred in the rainy season. The results indicate that fall in DO and rise in free CO_2 during rainy season could be ascribed to retarded photosynthetic activity of the phytoplanktons or decreased concentration of O_2 being consumed by the organic matter in turbid state of water during low phytoplankton density.

The adverse effect of turbidity on phytoplanktons may be due to the blanketing effect of the suspended material which interferes with photosynthetic activity (Welch, 1952).

Among zooplanktons, Michael (1964), Arora (1966) and Sampath *et al.* (1979) found some correlation between rotifers and chloride. Many workers showed a positive correlation between rotifers and temperature (Elliot 1976, Sampath *et al.*, 1979, Patil *et al.*, 1985). Michael and Sampath *et al.*(1979) observed an increase in the number of rotifers with hardness and alkalinity. A Positive correlation was also observed, in Tighra reservoir, between rotifers and chloride. Besides chloride, rotifers also showed positive correlation with alkalinity, hardness, temperature and pH. Negative correlation was observed between transparency and copepods.

Factors like low pH, temperature and alkalinity influence the occurrence of rotifers. (Singhai *et al.*, 1989; Chauhan, 1993; Puspendra and Madhyastha, 1994 and Bais and Agarwal, 1995).

Though temperature is not a limiting factor for the growth of zooplanktons, it plays an important role either directly or indirectly in their growth and production (Nasar 1977; Kumar, 1995; Rao *et al.*, 1996).

Mishra (2005) found a positive correlation between rotifers and hardness of water and a negative correlation between rotifers and water temperature. He observed a high positive correlation between zooplankton and phytoplankton populations. In addition, he observed a high positive correlation between zooplankton and phytoplanktons.

In Tighra reservoir, hardness showed a positive correlation with zooplanktons a well as with phytoplanktons. Among the zooplanktons, a highly positive correlation was found with rotifers and copepods.

An inverse relationship between population density of rotifers and the water current, as well as a direct relationship between the water temperature and the population density of the rotifers was observed by Shayestenfar *et al.* (2008).

Rajshekhar *et al.* (2010) observed correlation between zooplankton and physico-chemical parameters. High incidences of rotifers, during summer season, indicate the influence of temperature and a positive correlation between temperature and rotifer population.

Basu *et al.* (2010) found a positive relation between zooplankton abundance and pH, DO, combined CO_2, nitrate and phosphate.

In addition to the population of producers, the capacity of synthesizing organic matter by an aquatic ecosystem depends largely on physico-chemical parameters like temperature, nutrients and solar radiation. The wide fluctuations in physico-chemical features also impart considerable seasonal variation in the rate of primary productivity in the reservoirs (Ramkrishnaiah and Sarkar, 1982). Many workers have studied correlation between physico-chemical factors and productivity (Sreenivasan, 1964; Singh, 1990; Agarwal *et al.*, 1990; Kumar, 1996b; Prakasam and Joseph, 2000; Mali and Gajaria, 2004; Raghuwansi, 2005; Sivakumar and Karuppasamy, 2008).

Likens (1975) observed a close relationship between temperature and productivity. Khare (2005) found a marked and significant correlation among plankton density and temperature, DO, phosphate and nitrate. Fasihuddin and Puttaiah (2005) also found a positive correlation between temperature and productivity. They observed high production rate during pre-monsoon at high temperature. Shiddamallayya and Pratima (2007) also reported a positive correlation of GPP with atmospheric temperature and DO in Papnosh pond.

A highly positive relationship was also observed between productivity and temperature in the present study. Productivity also had positive correlation with pH, chloride, hardness, CO_2 and turbidity. A negative correlation was observed among transparency, GPP and NPP.

Das (2000) observed a significant relationship between seasonal variations in alkalinity with gross productivity and net productivity. Negative relation with alkalinity and NPP was recorded by Mahadev *et al.*(2003). Fasihuddin and Puttaiah (2005) found a strong positive relationship between alkalinity and productivity. But a positive relationship was observed between alkalinity and NPP in the present investigation.

Das (2000) analyzed correlation between transparency and primary production in some reservoirs of Andhra Pradesh. Kumar and Bohra (2005) found a negative correlation between productivity and secchi transparency and positive correlation between productivity and temperature. Fasihuddin and Puttaiah (2005) found a strong negative relationship between transparency and primary productivity. According to Bhat and Pandit (2005), higher transparency and temperature, associated with low water levels, seem to be conducive factors for maximum phytoplankton density.

Several studies indicate that nutrients also influence the rate of productivity. Sabu and Azis (1995) reported a strong positive relationship between phosphorous and productivity. Singh (2000) and Eshwaralal and Angadi (2002) found a positive relation between CR and phosphate, hardness and alkalinity. They recorded a negative relation of CR with NPP.

Eshwaralal and Angadi (2002) also reported a negative relation between NPP and phosphate. A close relationship of CR with CO_2, hardness, temperature and pH was observed in the present study.

Fasihuddin and Puttaiah (2005) reported a strong positive relationship between nitrates and phosphates with gross primary productivity. Bhat and Pandit (2005) reported a direct impact of concentration of phosphorus and nitrogen on the primary production and development of phytoplankton community.

Mahadev and Hosmani (2005) reported that higher pH, calcium and oxidizable organic matter coupled with low concentration of nitrate and phosphorus were found to be favoring the growth of diatoms. Similar findings were reported by Shiddamallayya and Pratima (2007). According to them phosphate is known to act as nutrient source of growth and production of phytoplanktons. They found that NPP positively fluctuate with GPP and showed a negative relation with phosphate. A positive relationship was also recorded between GPP and NPP in the present study.

Singh and Laura (2012) found a significant correlation between phytoplankton density and pH, magnesium hardness and total nitrogen ($P= 0.05$), total alkalinity, total hardness, calcium hardness, total dissolved solids and net primary productivity ($P= 0.01$).

Conclusion

Limnological studies of Tighra reservoir indicate that,in addition to physico-chemical factors, there are several environmental factors such as local hydrography, seasonal variability and biological parameters which act in regulating various properties and the productivity of a water body.

Physico-chemical factors like temperature, DO, hardness, phosphate, alkalinity and organic matter are responsible in regulating phytoplankton production and in determining the productivity of a water body. These parameters also influence the distribution of zooplanktons.

The present study reveals that physico-chemical factors not only influence biological parameters but these factors are also correlated with each other. The study could be helpful in better understanding of Tighra reservoir. On the basis of the present study of physico-chemical and biological parameters, it can be concluded that the water of Tighra Reservoir is safe for drinking purposes. In addition, the study indicates that the water is quite suitable for fish culture and by adopting new and better management and scientific approach as the production of the fish can be augmented. There is a lot of scope for the better utilization of the reservoir for aquaculture.

7

Summary

Limnology is a branch of science which deals with the study of freshwater ecosystems. According to Odum, studies on freshwater habitat with special reference to its physico-chemical, geological and biological characteristics is termed as limnology. Freshwater ecosystems are very important for human beings. Besides their use for drinking, house hold and industrial requirements; the lakes and reservoir are used for fish culture, aquaculture, irrigation, hydropower and various other purposes. Limnological studies of the freshwater bodies not only provide knowledge about the physico-chemical parameters of the water body but also enable us to explore better possibilities for augmentation of pisciculture.

Tighra reservoir is situated about 20 km west of Gwalior, near Tighra village which is in close proximity of SADA Magnet city. It lies on 26°13′ N latitude and 78° 30′ E longitude, at an altitude of 218. 58 m. Tighra reservoir, the life line of Gwalior, was primarily constructed to fulfill the water supply of the city. It is also used for irrigation purpose Water is also used to culture fish by the fisheries department.

Methodology

The water quality of a water body depends on its physico-chemical characteristics. Physico-chemical factors influence the aquatic flora and fauna, which in turn, also influence the physico-chemical properties of water.

The climatic conditions of the area and its geographical location also influence the physico-chemical and biological properties of water. Thus, the quality of the water of a water body depends on a number of factors. To assess the water quality of a water body, determination of physico-chemical, biological and climatic conditions all are important.

Limnological studies of Tighra reservoir were carried out for the period November 2010 to October 2011.

To determine the climatic parameters- air, temperature, rainfall and humidity, for each month, data were obtained from Metrological Department, Gwalior.

For physico-chemical and biological studies of the Tighra reservoir, the reservoir was divided into four zones. In each zone one sampling station was selected. The sampling stations were so selected as to cover the maximum area of the reservoir.

The water samples were collected monthly in the morning hours. The pH and temperature were estimated on the spot while other parameters were estimated in the laboratory.

For biological studies, samples were collected by filtering 50 liter surface water through plankton net The collected samples were preserved in lugol's solution for phytoplanktons and in 4 per cent formalin for zooplanktons. The preserved samples were brought to the laboratory for qualitative and quantitative analysis. Phytoplanktons were identified by using standard methods.

Quantitative studies were made by using Sedgwick rafter cell. All the planktons, present in the cell, were counted by moving the cell vertically and horizontally, covering the whole area.

Primary productivity was measured by using *Gaarder and Gran's method* of light and dark bottle.

Observation and Results

Climate Studies

The average minimum air temperature varied from 6.37 to 27.71 °C. The lowest temperature was recorded during January 2011 while the highest average minimum temperature was recorded in May 2011. Total monthly rainfall ranged between 0 (nil) to 340.5 mm.

Humidity was recorded at 8.30 AM and 5.30 PM daily by the Metrological Department. Data of the average humidity for the months were noted down. Average monthly humidity at 8.30 AM, which was the time close to the time of collection of the water samples, ranged between 42.45 per cent and 90.29 per cent. The average maximum humidity 90.29 per cent was recorded during January while humidity was minimum (42.45 per cent) during May 2011.

Physical Parameters

The average water temperature varied from 18.4 °C to 35.75 °C during the study period. The conductivityof the water of Tighra reservoir was recorded in the range of 272.5 µS/cm to 408.5 µS/cm.Transparency of the water was in the range of 152.75 cm to 211.5 cm.Turbidity of Tighra reservoir fell in the range of 5.77 NTU to 12.15 NTU.

Chemical Parameters

The average pH of the Tighra reservoir varied from 6.85 to 7.72. Dissolved oxygen of the reservoir was in the range of 5.425 to 8.125 mg/lit. Free CO_2 of Tighra reservoir was recorded between 4.15 mg/lit to 7.825 mg/lit.Total hardness of Tighra reservoir was between 66.25 mg/lit and 137 mg/lit. Water of Tighra reservoir showed alkalinity in the range of 53.75 mg/lit to 145.5 mg/lit). Chloride was recorded between 11.85 mg/lit and 39.5 mg/lit. Nitrate content was low in Tighra reservoir throughout the study. The values varied from 0.29 mg/lit to 0.61 mg/lit. The phosphate content of the Tighra reservoir was found in the range of 0.37 mg/lit to 1.57 mg/lit.

Seasonal variations were observed in various physico-chemical parameters.

Biological Studies

Zooplankton population of the reservoir composed of five major groups namely Rotifera, Copepoda, Cladocera and Protozoa. Total 20 species of zooplanktons were identified. Out of which 10 belonged to rotifera, 4 belonged to Copepoda, 4 to Cladocera and 2 were from Protozoa.

Four major groups of phytoplanktons *i.e.*, Bacillariophyceae, Myxophyceae, Chlorophyceae and Euglenophyceae were recorded from the reservoir during the study period. Out of the total 26 species of phytoplanktons, 9 belonged to Bacillariophyceae, 9 to Myxophyceae, 6 to

Chlorophyceae and 2 were from Euglenophyceae. Seasonal variations were also recorded in the distribution of phytoplanktons and zooplanktons.

Primary productivity was higher during summer than winter and other months of the study period.

Correlation between physico-chemical and biological factors was determined. Many of these factors found to be closely related to each other and influence each other.

Physico-chemical and biological studies of the water indicate that the water is quite suitable for drinking purpose and pisciculture.

References

Abbasi S.A., K.S. Bhatiya, A.V.M.Kunhi and R.S. Soni (1996). Studies on the Limnology of the Kuttadi Lake (North Kerala). *Eco. Env. Cons.* **2:** 17-27.

Achionye-Nzeh, Chioma G. and A. Isimaikaiye (2010). Fauna and Flora composition and water quality of a reservoir in Ilorin, Nigeria. *Int. J. of Lakes and Rivers* **3(1).** : 7-15.

Adebisi B.A. (1981). The physico-chemical hydrobiology of tropical river upper urgan river, Nigeria, *Hydrobio.* **79(2):** 757-765.

Adesalu T.A. and D.I. Nwankwo (2005). Studies on the phytoplanktons of Olero Creek and Parts of Benin River, Nigeria. In: Ecology of Planktons. Ed. Arvind Kumar. Daya Publishing House, Delhi pp.15-29.

Adholia U.N. and A.Vyas (1992). Correlation between copepods and limnochemistry of Mansarovar reservoir, Bhopal. *J.Env. Biol.* **13(4):** 281-290.

Adholia U.N. (1979). Zooplanktons of river Betwa, India. *Geo-Eco. Trop.* **3:** 267-271.

Adholia U.N. (1992). Bioenergetics of Mansarover Reservoir, Bhopal with reference to fish Production. Technical Report. Department of Zoology. Govt. Motilal Vigyan Mahavidyalaya, Bhopal.

Adoni A.D. (1985). Work Book on Limnology. Pritibha Publication, Sagar (MP) India.

Agarkar S.V., A.B. Bhosle and P.M.Patil (1998). Physico-chemical analysis of drinking water from Buldhana district, Maharashtra. *J.Aqua. Biol.* **13(1 and 2):** 62-63.

Agarwal Ashok K. and Govind S. Rajwar (2010). Physico-chemical and microbiological study of Tehri Dam Reservoir, Garhwal Himalaya, India. *J. of American Sci.* **6(6):** 65-71.

Agarwal N.C., V.S. Bais and S.N. Shukla (1990). Effect of nitrates and phosphate enrichment on primary productivity in the Sagar Lake, Sagar. *Poll. Res.* **9**: 29-32.

Agarwal N.K. and B.L. Thapliyal (2005). Pre-impoundment hydrological study of Bhilangana River from Tehri Dam reservoir area in Uttarakhand. *Environmental Geochemistry* **8:** 143-148.

Aher N.H. and S. N. Nandan (2005). Limnological studies of Mosam River in Maharashtra with relation to phytoplanktons. In : Ecology of Planktons. Ed. Arvind Kumar Daya Publishing House, Delhi pp.273-378.

Ahirrao S.D. and S.S. Pedge (2011). Water quality of Tridhara: a Holy place in Parbhani District, Maharashtra. Limnology Current Perspective (Ed. V.B. Sakhare). Daya Publisging House, Delhi pp.14-17.

Akin-oriola G.A. (2003). Zooplankton association and environmental factors in Ogupa and Ona rivers, Nigeria. *Rev. Biol. Trop.***51(2):** 391-398.

Alam J.B., A. Hossain, S.K. Khan, B.K. Banik, M.R. Islam, Z. Muyen and M.H. Rahman (2007). Deterioration of water quality of Surma River. *Environ. Monit. Assess.* **134(1-3):** 233-242.

Ali M.B., R.D.Tripathi, U.N.Rai, A. Pal and S.P.Singh (1999). Physico-chemical characteristics and pollution level of Lake Nainital (U.P., India): role of macrophytes and phytoplanktons in biomonitering and phytoremediation of toxic metal ions. *Chemosphere* **39(12)**: 2171-2182.

Angadi S.B., N. Shiddamallayya and P.C. Patil (2005). Limnological studies of Papnash Pond, Bidar (Karnataka). *J. Environ. Biol.* **26(2):** 213-216.

APHA (1989). Standard Methods for the Examination of water and waste water 17th edition. American Public Health Association, Washington D.C.

Arce R.B. and C.E. Boyd (1980). Water chemistry of Alabaena Ponds. Aubrn. Univ.(Ala). *Agri. Eco. Sta. Bull.* **322**: 35.

Arora H.C. (1966). Responses of rotifer to variations in some ecological factors. *Proc. Indian Acad.Sci.* **63**: 57-66.

Ayoade A.A., N.K. Agarwal and A. Chandela Solanki (2009). Changes in physic-chemical features and plankton of two regulated high altitude rivers Garwal Himalaya, India. *Euro. J. of Sci. Res.* **27(1):** 77-92.

Ayyappan S. and T.R.C. Gupta (1985). Limnology of Ramasamudra Tank Primary production. *Bull. Bot. Soc. Sagar.* **32:** 82-88.

Aznar R., C. Amaro, E. Garay, E. Alcaide (1991). Physico-chemical and bacteriological parameters in a hypereutrophic lagoon (Albafera Lake, Valencia, Spain). *Zentralbi Mikrobiol.* **146(4)**: 311-321.

Bade B.B., D.A.Kulkarni and A.C. Kumbhar (2009). Studies on physico–chemical parameters in Sai reservoir, Latur district, Maharashtra.*Int. Res. Jour.***7(II):** 31-34.

Badola S.P. and H.R. Singh (1981). Hydrology of the river Alaknanda of Garhwal Himalya, India. *J. Ecol.* **8:** 269-276.

Bais V.S. and N.C. Aggarwal (1995). Comparative study of the zooplanktonic spectrum in the Sagar Lake and military engineering Lake. *J. Environ. Biol.* **16(1):** 27-32.

Bais V.S., N.C. Aggarwal and A. Tazeen (1997). Seasonal changes in phytoplankton productivity due to artificial environment of nutrient. A situ experiment. *J. Environ. Biol.* **18(3):** 249-255.

Bais V.S., S. Yatheesh and N.C.Agarwal (1993). Evaluation of the trophic status of a lake. *Proceedings of the Academy of Environmental Biology.* **2:** 125-131.

Balkhi M.H. (1987). Hydrobiology of Anchor Lake, Kashmir. *Physiol. Eco.* **12(3):** 131-139.

Banerjee Ruma and Apurba Ratan Ghosh (2012). Assessment of physicochemical attributes of river Damodar from Barakar its upstream zone to Burdwan the downstream zone. *Int. Jr.of Current Research* **4(05):** 26-31.

Basavarajappa S.H., N.S. Raju, S.P. Hosmani and S.R. Niranjana (2010). Algal diversity and physico-chemical parameters in Hadhinam Lake, Mysore, Karnataka state, India. *The Bioscan* **5(3):** 377-382.

Basu M., N. Roy and A. Barik (2010). Seasonal abundance of Net Zooplankton correlated with physico-chemical parameters in a freshwater ecosystem. *Int. J. of Lakes and Rivers.* **3(1):** 67-77.

Battish S.K. (1992). Freshwater zooplankton of India. Oxford and IBH Publishing Co. Pvt. Ltd. New Delhi.

Beyst B., D. Buysse, A. Dewicke and J. Mees (2001). Surf zone hyperbenthos of Belgian sandy beaches: Seasonal patterns, Estuarine Coastal Shelf. *Sci.* **53**: 877-895.

Bhadra B., S.Mukherjee, R Chakraborty, A.K.Nanda (2003). Physico-chemical and bacteriological investigation in the river Torsa of North Bengal. *J. Environ.Bio.* **14(2)**: 125-133.

Bhagde Rupendra V. (2005). Study of physico-chemical parameters of Bhatye Estuary on Ratnagiri Coast of MS. *J. Aqua. Bio*. **20(2):** 113-116.

Bharti D.P.S and K.S. Rana (1987). Zooplankton in relation to biotic components in the Fortmoat of Bharatpur. *Proc. Nat. Acad. Sci. India.***57(B)**. : 237-242.

Bhat Mohd. Muzamil, Taiyyaba Yazdani, Kamini Narain, Mohammad Yunus and Ravinder Nath Shukla (2009). Water quality status of some urban ponds of Lucknow, Uttar Pradesh. *J. of Wetland Ecology*. **2**: 67-73.

Bhat Samin A. and Ashok K. Pandit (2005). Phytoplankton Dynamics in Anchar Lake, Kashmir. In : Ecology of Planktons. Ed. Arvind Kumar Daya Publisging House, Delhi pp.190-208.

Bhatt S.D. and Usha Negi (1985). Physico-chemical features of phytoplankton population in a subtropical pond. *Comp. Physiol. Ecol.* **10(2):** 85-88.

Bhosale Leela J. Surekha N. Dhumal and Anjali B. Sabale (2010). Seasonal variations in occurrence of phytoplankton and primary productivity of some selected Lakes in Maharashtra. *The Bioscan*. **2:** 569-578.

Bilgrami K.S., K.Sheo and S.Kumar (1994). Dial variations in abiotic and biotic factors of the river Ganga at Bhagalpur. *Environment and Ecology* **12:** 374-378.

Bohra O.P. (1977). Observation on the diel cycle of abiotic paramterss at Jatabera, Jodhpur. *Comp. Physiol. Ecol.* **2(3)**: 115-118.

Bohra C. and A. Kumar (1999). Comparative studies of phytoplankton in two ecologically different lentic freshwater ecosystems. Modern trends in environmental pollution and ecoplanning (Ed. A. Kumar). ABP Publishers, Jaipur pp.220-242.

Bohra C. and A. Kumar (2002). Primary productivity of a sewage fed aquatic ecosystem. Ecology and Ethology Biota Vol. I (Ed. A. Kumar). Daya Publishing House, Delhi, pp.373-392.

Buktar P.P. and V.B. Sakhare (2011). Studies on physico-chemical parameters of Wan reservoir on Beed District, Maharashtra. Limnology Current Perspective (Ed. V. B. Sakhare). Daya Publishing House, Delhi pp.18-22.

Cabecadas G. and M.J. Brogueira (1987). Primary production and pigments in three low alkalinity connected reservoirs receiving mine wastes. *Hydrobiol.* **144:** 173-182.

Chacko P.M. and P. I. Krishnamurthy (1954). On the plankton of three freshwater Fish ponds in Madras city, India. *Indo Pacific Fish Coun.* (UNESCO), pp 103-107.

Chakarabarty R.D., P. Roy and S.B. Singh(1957). A qualitative study of the plankton and physico-chemical conditions of the river Jamuna at Allahabad in 1954-55. *Indian J Fish* **6(1)**. : 186-203.

Chandler D.C. (1942). Limnological studies of Western lake Erie III. Phytoplankton and physical chemical data from November 1939 to November 1940. *Ohio J. Sci.* **42**: 24-44.

Chandrashekhar S.V.A.(1996). Ecological studies on Sorrornagar Lake, Hyderabad. Ph.D Thesis submitted to Osmania University, Hyderabad.

Chattopadhyay C. and A. Barik (2009). The composition and diversity of Net zooplankton species in a tropical freshwater lake. *Int. J. of Lakes and Rivers.* **2(1):** 21-30.

Chauhan Ramesh (1993). Seasonal fluctuations of zooplankton on Renuka Lake, Himachal Pradesh, UP. *J. Zool.* **13(1):** 17-20.

Chavan R.J., A.D. Mohekar, R.J. Savant and M.B. Tat (2005). Seasonal variations of abiotic factors of Manjra Project Water Reservoir in District Beed, Maharashtra, India. *Poll. Res.* **24(3):** 705-708.

Chopra A.K. and Patric Nirma J. (1994). Effect of domestic sewage on self purification of Ganga water at Rishikesh I. Physico-chemical parameters. *Ad. Bios.* **13(11):** 75-82.

Choudhury S.K., U. Bhattacharya and S.K. Sinha (1987)Perodicity of phytoplanktons in freshwater ponds at Darbhanga.*Environ. and Ecol.* **5(3):** 223-241.

Dagaonkar A. and D.N. Saksena (1992). Physico-chemical and biological characterization of a temple Tank, Kaila Sagar, Gwalior(M.P.). *J. Hydrobiol.* **VIII(1):** 1-19.

Das A.K. (2002). Phytoplankton primary production in some selected reservoirs of A.P. *Geobios.* **29:** 52-57.

Davis G.L.(1955). The marine and freshwater plankton in Michigan state. University Press, East Lansing.

Deshmukh U.S. (2001). Ecological studies of Chhatri Lake, Amravati with special reference to plankton and productivity. Ph.D. thesis Amravati University, Amravati.

Deswal S. and P. Chandana (2007). Water quality status of Sannihit Sarowar in Kurukshetra (India). *J.Environ Sci Eng.* **40(1):** 51-563.

Dhanapathi, M.V.S.S.S. (2000). Taxonomic notes on the rotifers from India (from 1889-2000). Indian Association of Aquatic Biologists Publ. No.10.

Dhar Sheetu and Deepika Slathia(2014). Seasonal Water Quality Assessment of Lake Mansar, Jammu, J and K (India). *IJSR* **3(7):** 215-217.

Dutta N.C., B.K. Banerjee and S.B.Banerjee (1983). Relationship between phytoplankton and primary productivity of a freshwater pond in Calcutta. *Environment and Ecology* **1:** 135-137.

Dutta N.C., B.R. Banerjee and N.K.Dar (1982). Diurnal rhythm of physicochemical properties and zooplankton in a freshwater pond in Calcutta. Ind. *J. Physiol. and Nat. Sci.* **2**; 22-27.

Dutta S.P.S., H. Kumar and J.P.S. Bali (2001). Hydrological studies on River Basantan Samba, Jammu (J and K). *J. Aqua. Biol.* **16(1):** 41-44.

Elliot I.J.(1976). Seasonal changes in the abundance and distribution of planktonic rotifers in Grasmere (English lake District). *Freshwater Biol.* **8:** 144-166.

Ellis M.M. (1937). Detection and measurement of stream pollution. *U.S. Bur. Fish. Bull.* **22**: 367-437.

Elmaci A., F.O. Topac, N. Ozengin, A. Teksoy, S. Kurtoglu and H.S. Baskaya (2008). Evaluation of physical, chemical and microbiological properties of lake Uluabat Turkey. *J. Environ. Biol.* **29(2):** 205-210.

Escribano R. and P. Hidalgo (2000). Spatial distribution of copepods in the North of the Humboldt Current region of Chile during coastal upwelling. *J.Mar.Biol.Assoc.UK* **80:** 283-290.

Eshwaralal S. and S.B. Angali (2002). Primary productivity of two freshwater bodies of Gulbarga. India. *Nat. Environ. And Poll. Tech* **1(2):** 151-157.

Eswarlakshmi R., J. Jayanthi and H.G. Ragunathan (2008). Physico-chemical factors of the water of Vattambakkam Lake in Kanchipuram District, Tamil Nadu, India. In: Advances in aquatic ecology vol. II Ed. V.B. Sakhare. Daya Publishing House, Delhi pp. 65-68.

Fasihuddin Syed and E.T. Puttaiah (2005). Spacio-temporal trends in phytoplankton Primary Productivity in Manikanava Reservoir. In: Ecology of Planktons. Ed. Arvind Kumar. Daya Publishing House, Delhi. pp. 127-134.

Ferdous Zannatul and A.K.M. Muktadir (2009). A review: Potentiality of Zooplankton as Bioindicator. *Am. J. Applied Sci.* **6(10):** 1815-1819.

Franco David A. (1987). Limnological characteristics of Sangre Lake, Okalhoma (USA).*Jr. of Freshwater Ecol.* **4 (1):** 53-60.

Gadded S.M., P.M. Nimbargi and S.S. Rodgi (1983). Ecological studies of two polluted water bodies of Gulbarga. *Poll. Res.* 2(2): 49-52.

Ganapti S.V. (1940). The ecology of a temple tank containing a permanent bloom of *microcystis aeruginosa* (Kutz). Henfr. *J.Bombay Nat. Hist. Soc.* **42:** 65-67.

Ganapti S.V. (1949). An ecological study of a pond containing zooplankton. *Proceedings of Indian Academy of Science* **17:** 41-58.

Ganapti S.V. (1950). Suggestions for stocking fish ponds in Madras. *Madras Agric.* **37**: 169-173.

Ganapti S.V. (1956). Hydrobiological investigations of the Hope reservoir of Thmbaraparani river at Papanasam Tirunalveli Dt. Madras State. *Indian Geogr. J.***31:** 1-2.

Ganapti S.V. (1960). The ecology of tropical waters. *Proc. Symposium on Algology*, 1959 (ICAR 1960). pp. 204-218.

Garg R.K., R. J. Rao, D. Uchchariya, G. Shukla and D.N. Saksena (2010). Seasonal variations on water quality and major threats to Ram sagar reservoir, India. *Afr. J. Environ. Sci. Technol.* **4(2):** 61-76.

George M.G. (1961). Observation on the rotifer from shallow ponds in Delhi. *Curr.Sci.30:* 265-269.

George M.G. (1962). Occurrence of permanent algal bloom in a Fish tank at Delhi with special reference to factors responsible for its production. *Proc.Indian Acad. Sci.* **56(8)**. : 354-362.

Goel P.K and V.B. Autade (1995). Zoological studies in the river Panchganga at Kolhapur with emphasis on biological components. In: Recent Research in aquatic environment. Ed. V.B. Ashutosh Goutam and N.K.Aggarwal. Daya Publishing House, Delhi pp. 25-46.

Golterman H.L. (1969). Methods for chemical analysis of waste water. IBP Hand book.

Golterman H.L. (1975). Physiological limnology.Elsevier Scientific Pub.Co. NY.

Gonjari G.R. and R.B. Patil (2008). Hydrobiological studies on Triputi reservoir near Satara, Maharashtra. *J. Aqua.Bio.* **23(2):** 73-76.

Gopal B., K.P.Sharma and R.K. Trivedy (1978). Studies on ecology and production in Indian freshwater ecosystems at primary producer level with emphasis on macrophytes. In: Glimpses of ecology Ed. J.S. Singh and B. Gopal. Int. Sci. Pub., Jaipur. pp.349-376.

Goswani S.C. (2004). Zooplankton, Methodology, Collection and Identification- a field manual. National Institute of Oceanography, Dona Paula, Goa.

Govind B.V. (1963). Preliminary studies on plankton of the Tungabhadra reservoir. *Indian J. Fish.* **(10):** 148-158.

Green J.(1960). Zooplankton of the river Sokoto: The Rotifera. *Proc. Zool. Soc.Lond.* **135:** 491-523.

Gurung Smriti, Subodh Sharma and Chatra Mani Sharma (2009). A brief review on limnological status of high altitude lakes in Nepal. *J. of Wetland Eco.* **3:** 12-22.

Harish Kumara B.K. and S. Srikantaswamy (2011). Seasonal variation of plankton diversity in Tungabhadra river of India. *Asian J. Environ. Sci.* **6(1):** 80-88.

Harsha T.S. and S.G. Malammanver (2004). Assessment of phytoplankton density in relation to environmnental variables on Gopalaswamy pond at Chitradurga, Karnataka. *J. Environ. Biol.* **25(1):** 113-116.

hppcb.gov.in/EIAsorang/Spec.pdf.

http: //wgbis.ces.iisc.ernetin/energy/water/paper/Tr-115/ref.htm.

Hulyal S.B. and B.B. Kaliwal (2008). Dynamics of phytoplankton on relation to physico-chemical factors of Almathi reservoir of Bijapur District, Karnataka state. *Environ. Monit. Assess.* **153(1-4):** 45-59.

Hulyal S.B. and B.B. Kaliwal (2009). Water quality assessment of Almatti Reservoir of Bijapur (Karnataka state, India). with special reference to zooplankton. *Environ. Monit. Assess.* **139(1-3):** 299-306.

Hussainy S.V. (1967). Studies on the limnogy and primary production of a tropical lake. *Hydrobiol.* **30(3)**. : 335-352.

Hutchinson G.E. (1941). Limnological studies in conna.IV Mechanism of intermediary metablosim in stratified lakes. *Ecol.Mongr.***11**: 21-50.

Hutchinson G.E. (1957). A treatise on Limnology Vol. Part 2 Chemistry of Lakes, John Wiley and Sons. Inc., USA.

Hynes H.B.N. (1970). The ecology of running waters. Liverpool University Press, Liverpool IV Impression.

Ingole S.B., G.A. Kadam, S.B. Naik and G.K. Kulkarni (2011). Water quality of Majalgaon Dam with special reference to Zooplankton. Limnology current Perspective Ed. V.B. Sakhare, Daya Publishing House, New Delhi. pp. 248-263.

ISI (1983). Indian Standard Specification for Drinking Water. ISI 10500.

Jagtap V.P., H.K. Bhagwan and S.M. Kamble (2011). Studies on seasonal variation on physico-chemical characteristics of Sina-Kolegaon Reservoir in Osmanabad District, Maharashtra. Limnology Current Perspective (Ed. V.B. Sakhare). Daya Publishing House, Delhi pp.23-31.

Jain Renu, D. K. Sharma and Dushyant Sharma (2002). Studies on the Ecology and Fish fauna of Gopalpura Tank of Guna District, Madhya Pradesh. *Environment Conservation Journal* **3(2)**. : 49-52.

Jain Renu and Dushyant Sharma (2000). Water quality of Rampur reservoir of Guna District, Madhya Pradesh. *Environment Conservation Journal* **1(2 and 3).** : 99-102.

Jana B.B. (1979). Temporal plankton succession and ecology of a subtropical Tank in West Bengal, India. *Int. Rev. Gas. Hydrobiol.* **64(5):** 661-671.

Jana B.B., V.K. De and R.N. Das (1978). Ecology and plankton production in two fish pond Kalyani, West Bengal.*Proc.65 India Sci.cong. Assoc. Abstract* 303-304.

Jayabhaye U M., M.S. Pentewar and C.J. Hiware (2008). A study on physico-chemical parameters of a minor reservoir Sawana, Hingoli District of Maharashtra. *J. Aqua. Biol.* **21(2).** : 56-60.

Jeelani M., H. Kaur and S.G. Sarwar (2005). Distribution and ecology of phytoplanktons in the Dal lake Kashmir (India). In: Ecology of Planktons. Ed. Arvind Kumar. Daya Publishing House, Delhi pp.41-54.

Jeyachandran A. and N. Krishnan (2001). Hydrobiology and primary productivity of Viraganur dam, Madurai, Tamil Nadu. Indian *Hydrobiol.* **4(1):** 39-42.

Jha Prithviraj and Sudip Barel (2003). Hydrobiological study of lake Mirik in Darjeling, Himalaya. *J. Envion. Bio.* **24(3):** 339-344.

Jhingran A.G. (1989). Limnology and production biology of two man-made lakes on Rajasthan (India). with management strategies for their fish yield optimization. Final Report IDA Fisheries Management in Rajasthan. Central Inland Fisheries Research Institute, Barrackpore, India. pp.1-63.

Jhingran Arun G. and V.V. Sugunan (1990). General guidelines and planning criteria for small reservoir fisheries management. Reservoir Fisheries in India (Ed. Jhingran A.G. and V.K. Unnithan). 1-8, Asian Fisheries Society, Indian Branch, Mangalore, India.

Joshi B.D., R.C.S. Bisht, Namita Joshi (1996). Planktonic population in relation to certain physico-chemical factors of Ganga Canal at Jwalapur (Haridwar). Him. *J. Env. Zool.* **10:** 75-77.

Joshi C.B. (1996). Hydrobiological profile of river Sutlej in its middle stretch in Western Himalayas. *Uttar Pradesh J.Zool.* **16(2)**;97-103.

Joshi C.B.and B.C. Tyagi (1997). Some limnological features of a cold stream-The Chirapani in Kumaon Himalaya. *Uttar Pradesh J.Zool.* **17(2)**. : 156-162.

Joshi P.K. (2006). Physico-chemical analysis of water from Ekruk Reservoir, Maharashta for potability. In: Ecology of Lakes and Reservoirs. Ed. V.B. Sakhare, Daya publishing House, Delhi. pp.132-137.

Kadam M.S., D.V.Pampatwar and R.P.Mali (2007). Seasonal variations in different physico-chemical characteristics in Masoli Reservoir of Parbhani District (MS). *J.Aqua. Biol.* **22(1)**. : 110-112.

Kadam S.U., J.M. Gaikwad and Md. Babar (2006). Water quality and ecological studies of Masoli Reservoir in Parbhani District, Maharashtra. Ecology of Lakes and Reservoir Ed. V.B. Sakhare, Daya Publishing House, Delhi. pp. 163-175.

Kamal D., A.N. Khan, M.A. Rahman and F. Ahamad (2007). Study on the physico-chemical properties of water of Mouri River, Khulna, Bangladesh. *Pak. J. Biol. Sci.* **10(5):** 710-717.

Kamble S.P. and C.V Muley (2008). Study on some physico-chemical parameters of Kalbadevi estuary in Ratnagiri District of Maharasthra. *J. Aqua. Biol.* **23(2):** 61-66.

Kargul R., A. Samandar, M. Yilmaz, L. Altun and R. Gedikli (2005). Evaluating the seasonal changes of some water quality parameters of the Buyuk Melon River Basin (Duzce, Turken). *J. Environ. Biol.* **26(2):** 179-185.

Kartha K.N. and K.S. Rao (1992). Environmental studies of Gandhisagar reservoir. *Fishery Techno* **29(1):** 14-20.

Kaur H.,S.S. Ddhillon, K.S. bath and G. Mander (1999). Correlation of rotifers with physico-chemical factors at Harika Reservoir (Punjab-India). In: Freshwater Ecosystems of India. Ed. Vijaykumar. Daya Publishing House, New Delhi pp. 187-192.

Kaushik K.S.,M.S. Agarkar and D.N. Saksena (1991a). Water quality and periodicity of phytoplantonic algae in Chambal Tal, Gwalior, Madhya Pradesh. *Bionature* **11(2):** 87-94.

Kaushik K.S.,M.S. Agarkar and D.N. Saksena (1991b). Distribution of phytoplankton in river water in Chambal area, M.P., *Bionature* **12:** 1-7.

Kaushik S.,M.N. Saksena and D.N. Saksena (1991c). Phytoplankton population dynamics in relation to environmental parameters in Matsya Sarovar at Gwalior (MP). *Acta. Botanica.* **19:** 113-119.

Kaushik S. and D.N. Saksena (1999). Physico-chemical limnology of certain water bodies of central India. In Freshwater ecosystem of India. Ed. K. Vijayakumar, Daya Publishing House, Delhi. pp 1-58.

Khaire B.S., A.D. Mohekar and R.J. Chavan (2011). Assessment of physico-chemical parameters and fish diversity of Mehakari Reservoir in Beed District, Maharashtra. Limnology current Perspective Ed. V.B. Sakhare, Daya Publishing House, New Delhi pp. 201-206.

Khan A.A. and A.R. Siddiqui (1971). Primary production in a tropical fish pond at Aligarh, India. *Hydrobiol.***37:** 447-456.

Khan I.A., A.A. khan and N. Haque (1988). Primary production in Sheikh jheel at Aligarh. *Environ. Ecol.* **6:** 858-862.

Khanna D.R. and R. Bhutani (2008). Laboratory manual of water and waste water analysis. Daya Publishing House, Delhi.

Khanna D.R. and R. K. Singh (2002). Seasonal fluctuations in the planktons of Suswa River at Raiwala (Dehradun). *Environment Conservation Journal* **1(2 and 3).** : 89-92.

Khanna D.R. (1993). Ecology and pollution of Ganga River, Ashish Publishing House, New Delhi.

Khanna D.R.,Shakun Singh, Ashutosh Gautam and J. P. Singh (2003). Assessment of water quality of river Ganga in District Bulandshahar (U.P.). India *J.Natcon* **15(1):** 167-175.

Khare P.K. (2005). Physico-chemical characteristics in relation to Abundance of plankton of Jagat Sagar Pond, Chattapur, India. Advances in Limnology Ed. S.R. Mishra, Daya Publishing House, New Delhi pp.162-174.

Kohli M.P. (1981). Plankton study of Govind sagar reservoir (India). *Comp. Physiol. Ecol.***6 (1):** 49-52.

Kulshrestha S.K., U.N. Adholia, A.Bhatnagar, A.A. Khan, M.Saxena and M. Baghail (1989). Studies on pollution in river Kshipra: Zooplankton in relation to water quality. *Int. J. Ecol. Environ.Sci.* **15:** 27-36.

Kumar A. (1995). Some Limnological aspects of the freshwater tropical wetland of Santhal Pargana (Bihar), India. *J. Environ and Poll.* **2(3):** 131-141.

Kumar A. (1996a). Impact of coal mining effluent on the primary productivity of water bodies and farming in and around the Chitra Coal fields in Santal Pargana, Bihar. India Environmental Management in Coal Mining and Thermal Power Plants (Ed. P.C. Mishra and Naik). Technoscience Publi. Jaipur, pp.188-196.

Kumar A. (1996b). Impact of sewage pollution on chemistry and primary productivity of two freshwater bodies in Santal Pargana (Bihar). *Indian J. Ecol.* **23**: 86-92.

Kumar A. and C. Bohra (2005). Dynamics of phytoplankton productivity of certain Lentic Ecosystem of Jharkhand, India. Ecology of plankton (Ed. Arvind Kumar). Daya Publishing House, Delhi pp. 1-14.

Kumar Arvind (1995). Periodicity and abundance of plankton in relation to physico-chemical characteristics of a tropical wetland of South Bihar. *Eco. Env. and Cons.* **1(1-4)**: 47-51.

Kundu S.K. and S.K.Sarkar (1987). Studies on zooplankton with reference to copepods in river water. *Proceedings of National Seminar on Estuary Management, Trivandram.* 4-5 June, 323-326.

Lashkar H.S. and S. Gupta (2009). Phytoplankton diversity and dynamics of Chatla flood plain Lake, Borak Valley, Assam, North East India- a seasonal study. *J. Environ. Biol.* **30(6):** 1007-1012.

Laxminarayana J.S. (1965). On the phytoplanktons of the river Gaqnga, Varanasi, India. *Hydrobiolgia* **25**: 119-137.

Liken G.E. (1975). Primary production of inland aquatic ecosystems. In: The primary productivity of Biospheres. Ed. Lieth and R.H.Whittakar, Springer-Verlag. New York.

Lokhande M.V., D.S. Rathod, V.S. Shembekar and K.G. Dande (2010). Studies on oxygen levels and temperature fluctuations in Dhanegaon Reservoir in Osmanabad. *Advances on Aquatic Ecology* Vol.**3** (Ed. V. B. Sakhare). Daya Publisging House, Delhi pp. 152-157.

Mahadev J., Nagarathnamma and Hosmani S. 2003 The significance of cluster analysis in determining relationship between physico-chemical parameters and phytoplanktons in polluted waters. *J.Eco.Toxicol.Environ. Monit.* **13(1):** 53-58.

Mahajan S.K., Yogesh Garde, Rakesh Bhawsar and Sharad Sharma (2002). Physico-chemical and biological characterization of the river Kunda at downstream of Khargaon Madhya Pradesh, India. *Environment Conservation Journal* **3(2)**. : 53-55.

Mali K.N. and S.C. Gajaria (2004). Assessment of primary productivity and hydrobiology characterization of a Fish culture pond, Gujarat. *Indian Hydrobiol.* **7(1 and 2)**: 113-119.

Mali R.P., A.T. Kamble and L.M. Mudkhede (2006). Physico-chemical analysis of Bhategaon Reservoir in Parbhani District, Maharashtra. Ecology of Lakes and Reservoir Ed. V.B. Sakhare, Daya publishing House, Delhi. pp.176-180.

Mane A.M. and S.K.Pawar (2007). Some physico-chemical properties of Manar rivear of Nanded District (MS). *J.Aqua. Biol.***22(2)**. : 88-90.

Mani Bharart and S.A. Gaikwad (1998). Physico-chemical characteristics of Lake Pokhran. *Indian J. Environ. And Toxicol.* **8(2):** 56-58.

Maniappa S. and V.K. Naik (2007). Physico-chemical properties of Malaprabha River. *J. Environ. Biol.* **49(1):** 1-6.

Mathew P. M. (1992). Seasonal primary productivity in the microphytic community of Govind garh lake, Rewa (M.P.). *J. Ecobiol.* **4(1):** 11-14.

Medina-Junior P.B. and A.C. Rietzler (2005). Limnological study of a Pantanal saline Lake Braz. *J. Biol.* **65(4)**: 651-659.

Mehra N.K. (1986). Studies on primary productivity in a subtropical lake: Comparison between experimental and periodicity values. *Indian J. Exp. Biol.* **24**: 189-192.

Michael R.G. (1968). Studies on the zooplanktons of a tropical fish pond. *Hydrobiolgia* **32**: 47-68.

Mishra G.P. and A.K.Yadav (1978). A comparative study of physico-chemical characteristics of lake and river water in central India. *Hydrobiol.* **59 (3):** 275-278.

Mishra S.R. (1993). Phytoplankton composition of sewage polluted Morar (Kalpi). river in Gwalior (M.P.). *Env. Eco.* **11(3)**: 625-629.

Mishra S.R. (2005). Zooplankton and their seasonal variation in a sewage collecting River at Gwalior, Madhya Predesh. In: Advances in Limnology. Ed. S.R. Mishra, Daya Publishing House, New Delhi pp.1-44.

Mishra Y.K. and B.K. Nayak (2000). Biomass productivity of energy plantations on Himalayan Hill. *Environ. and Ecol.* **18**: 13-16.

Moharana, Puspalata and A. K. Patra(2013). Primary Productivity of Bay of Bengalat Digha in West Bengal, India. *Indian J.L.Sci.* **3(1)**. : 129-132.

Muhamed Ashraf P. and M.K.Mukundan (2007). Seasonal variation in water quality of four stations in the Periyar river basins. *J. Environ. Sci. Eng.* **49(2)**: 127-132.

Mukherji Monika and N.C. Nandi (2006). Ecology Biodiversity and Management of Rabindar Sarovar in Kolkata, West Bengal. In : Ecology of Lakes and Reservoirs. Ed. V.B. Sakhare, Daya publishing House, Delhi. pp. 36-53.

Munawar M. (1970). Limnological studies of freshwater ponds of Hyderabad, India. *Hydrobiologia* **35**: 127-162.

Murugan N., P.Murrugavel and M.S. Kodarkar (1998). Cladocera: The biology, classification, identification and ecology. *Indian Association of Aquatic Biologists (IAAB).Hyderabad.*

Nasar S.A.K. (1977). Investigation in the seasonal periodicity of zooplankton on a freshwater pond in Bhagalpur. *Acta. Hydro. Chem. Hydrobid.* **5(6)**: 577-584.

Nayak T.R., J.Iyar and S.C.Jain (1982). Seasonal variation of rotifer and certain physico-chemical factors of Matyatal Panna.India *Comp.Physiol. Ecol.***7(3)**: 159-169.

Nogueina Marcos G., Paula Carolone Reis oliveira and yvana Tenorio de Britto (2008). Zooplankton assemblages (copepode and cladocera). in a cascade of reservoir of a large tropical river (S E Brazil). *Limnetica.* **27(1)**: 151-170.

Nygaard G. (1949). Studies on some pond and lakes II. The quotient hypothesis and some new or little known phytoplankton organisms. Danska Vidersk Selwsk **7**: 1-293.

Palaniappan N., R. Marimuthu and M. Rajendran (1981). Seasonal variation in primary productivity of two rock pools in Salem (TN), pp.58-72.

Palhaya J.P. and Malvia (1988). Pollution of the Narmada River at Hoshangabad in Madhya Pradesh and suggested measures for control. In: Ecology and pollution of Indian rivers.Ed. R.K.Trivedy. Ashish Publishing House, New Delhi. pp. 54-85.

Pandey B.C., B. K.Singh and P.N.Rao (1992). Species diversity of plankton in fish pond at Pune, Maharashtra. *Proceedings of Academy of Environmental Biology.* **1**: 167-172.

Pandey B.N., A.K. Mishra, P.K.L. Das and A.K. Jha (1995). Studies on hydrological conditions of river Saura in relation to its impact on Biological health. In: Recent Research in aquatic environment. Ed. V.B. Ashutosh Goutam and N.K.Aggarwal. Daya Publishing House, Delhi pp. 57-65.

Pandey Gajraj, S.N. Chaubey, Jayaraj Pandey and N.K. Shrivastava (2011). Water quality monitoring of river Tonsat, Akbarpur UP. *Indian J. Sci. Res.* **2(4):** 129-130.

Pandey J. and A. Verma (2004). The influence of catchment on chemical and biological characteristics of two feresh water tropical lakes of Southern Rajathan. *J.Environ.Biol.* **25**: 81-87.

Panigrahi, Subhashree and A.K.Patra(2013). Studies on seasonal variations in phytoplankton diversity of river Mahanadi, Cuttack city, Odisha, India. *Indian J.Sci.Res.* **4(2)**. : 211-217.

Parray S.Y., Sheerar Ahmad and S.M. Zubair (2010). Limnological profile of a sub urban wetland-chatlam Kashmir. *Int. J. of lakes and Rivers.* **3(1):** 1-6.

Patel Pushpa, S.K.Mahajan and S.K.Pathak (2002). Hydrobiological observations on Virla reservoir at Khargaon(Madhya Pradesh). *Environment Conservation Journal* **3(2)**. : 57-79.

Pati S. and Sahu B.K. 91993). Studies on primary production in Rengali reservoir (Odisha). *J. Ecobiol.* **5**: 85-88.

Patil S.G., D.K. Harshey and D.F. Singh (1985). Limnological studies of a tropical freshwater fish tank at Jabalpur M.P. *Geobios New Reports* **42(2):** 143-148.

Pawar S.K. and J.S. Pulle (2005). Study on physico-chemical parameters in Pethwadaj Dam, Nanded District of Maharashtra, India. *J. Aqua. Bio.* **20(2):** 123-128.

Pennak R.W. (1978). Freshwater Invertebrates of the United States. 2nd Ed. John Wiley Sons, New York, pp.803.

Philipose M.T. (1970). Freshwater plankton of Inland fisheries. *Pro. Sympo. Algae*. ICAR, New Delhi. pp. 272-291.

Poi A. and A.N. Shukla (1997). Studies on the reproductive biology of the cladoceran population of Gandhisagar reservoir, M.P., India. *Eco-development and Environment*.Ed.S P Singh pp: 137-141.

Pokale W.K.,J.N.Thakre and S.R. Warshate (2010). Water quality status of Pench Reservoir (India). *J.Environ Sci.Eng.* **52(3).** : 255-258.

Prakasam V.R. and M.L. Joseph (2000). Water quality of Sasthamcotta lake, Kerala (India). in relation to primary productivity and pollution from anthropogenic sources. *J. Environ Biol.* **21(4):** 305-307.

Prakash C., d.c Rawat and P.P. Grover (1978). Ecological study of river Jamuna.*IAWC Tech Annual* **5:** 32-45.

Prakash M.M. (1996). Correlation among physico-chemical characteristics of pond water (Mehata pond Jhabua). In: Assessment of water pollution. Ed. S R Mishra.A P H Publishing Corporation, New Delhi.pp 444-451.

Prakasad D.Y. and M.A. Qayyum (1978). Pollution aspects of upper lake of Bhopal.*Ind.Jour.Zool.* **4(1):** 35-46.

Pushpendra and M.N. Madhyastha (1994). Seasonal variation and diversity of zooplankton in a small pond near Manglore. *J. Ecobiol.***6(3):** 197-200.

Qadri M.Y. and A.R. Yousuf (1980). Limnological studies on Lake Malpur. *Geobios*. **7**: 117-119.

Quadri Syed Atheruddin(2015). Diversity of Planktons in Kham river, Aurangabad. IJSR **4(4):** 635-636.

Raghuwansi Arun K. (2005). The impact of physico-chemical parameter of lower lake, Bhopal on the Productivity of *Eichhornia carassipes.Eco. Env. and Cons.* **11(3-4):** 333-336.

Rai D.N. and U.P. Sharma (1991). Energy transfer efficiency and primary productivity of tropical wetlands in north Bihar (India). *J. Freshwater Biol.* **3**: 89-97.

Rai L.C. (1978). Ecological studies on algal communities of river Ganga at Varanasi *Ind. J. Ecol* **5**: 1-6.

Rajalakshmi S. and G. Krishnamoorthy (2007). Hydrological variations in mangroves of Puducherry India *J. Aqua. Biol.* **22(1):** 77-80.

Rajashekhar M., K. Vijaykumar and Zeba Paerveen (2010). Seasonal variations of Zooplankton community in freshwater reservoir Gulberga District, Karnataka, South India. *Int. J. of Systems. Biology.* **2(1):** 6-11.

Rajkumar N. (2004). A review on the quantum of phytoplanktonic primary production of polluted freshwater pond. *Indian Hydrobiol.* **7(1 and 2):** 61-65.

Ram P. and A.K. Singh (2007). Ganga water quality at Patna with references to physico-chemical and bacteriological parameters. *J. Environ. Sci. Eng.* **49(1):** 28-32.

Ramkrishnaiah M. and S.K. Sarkar (1982). Plankton productivity in relation to certain hydrological factors in Konar reservoir (Bihar). *J. Inland Fish. Soc. India* **14**: 58-68.

Rao C.B. (1955). On the distribution of algae in a group of six small ponds. II. Algae periodicity *J. Ecol.* **43:** 291-308.

Rao K.K., V.N. Rao and S.K. Muhmood (1996). Assessment of water quality and pollution of Narsingi Pond. *Eco. Env. and Cons.* **2**: 45-49.

Roy L.K., R.B. Gupta and V.P. Shrotry (2015). Seasonal variation in physico chemical characteristics of river Seep at Sheopur, India. In Emerging trends in Biological Sciences. Ed. Ram Pratap Singh and Vinayak Singh Tomar. Global book Organization, Delhi pp. 83-105.

Ruttner K. (1963). Fundamentals of limnology F. E. J. Univ. Press, Oxford University Press, Torento.

Sabu Thomas and P.K. Abdul Azis (1995). A study on the primary production in Peppara Reservoir. *Proc. of Seventh Kerala science congress. Palakkad.* pp. 76-77.

Saha L.C. and B. Pandit (1986). Comparative limnology of Bhagalpur ponds. *Comp. Physiol. Ecol.* **1(14)**: 213-216.

Saha L.C. and B. Pandit (1990). Dynamics of primary productivity between lentic and lotic systems in relation to biotic factors. *Ind. J. Bot. Society* **69**: 213-217.

Sahai, R and A. B. Sinha (1969). Investigation on Bioecology of inland of Gorkhpur (U.P.). India I. Limnology of Ramgarh Lake. *Hydrobioloia* **34(3-40)**. : 433-447.

Sahib,S.S. (2004). Physico-chemical parameters and zooplankton of the Shendurni River,Kerala. *J. Ecobiol.* **16** : 159-160.

Sakhare (2005). Water quality of Hingni (Pangaon). Reservoir and its significance to fisheries. Advances in Limnology Ed. S. R. Mishra, Daya Publishing House, New Delhi pp. 231-235.

Sakhare V.B. (2006). Ecology of Jawalgaon Reservoir in Solapur District, Maharashtra. Ecology of Lakes and Reservoir Ed. V.B. Sakhare, Daya publishing House, Delhi. pp.16-35.

Sakhare V.B. and P K Joshi (2003). Physico-chemical Limnology of Papnas- a minor Wetland in Tuljapur Town, Maharashtra. *J. Aqua. Bio.* **18(2):** 93-95.

Saksena D.N. and R.Mishra (1988). Characterization of wastes from industrial complex at Birla Nagar, Gwalior. *IAWPC Tech.Ann.* **15:** 163-167.

Saksena D.N. and N. Kulkarni (1986). On the rotifer fauna of two sewage channels of Gwalior (India). *Limnologica* **17:** 139-148.

Saksena D.N., R.K. Garg and R.J. Rao (2008). Water quality and pollution status of Chambal River in National Chambal Sanctuary, Madhya Pradesh. *J. Environ. Biol.* **29(5):** 701-710.

Salgotra Vikas, C.A. Mastan, Adarsh Kumar and T.A. Qureshi (2005). Studies on certain physico-chemical parameters of Hataikhede Reservoir near Bhopal, India. In: Advances in Aquatic Ecology Vol. I Ed. V.B. Sakhare, Daya Publishing House, New Delhi pp. 75-78.

Sanal Kumar M.G., Reeja Jose, Firozia Naseema Jalal and G. Santhy (2011). Seasonal variations in physico-chemical and microbiological properties of water of river Achencovil, Kerala. *J. Ecobiol.***28 (1)**: 95-100.

Sampath V, A. sreenivaan and A.Ananthnarayan (1979). Rotifers as biological indicators of water quality of Cauvery River. *Proc. Symp. On Environ. Biol.* Muzaffarnagar.

Sarkar S.K. and P.Basu Chowdhury (1999). Role of Some encvironmental factors on the Fluctuation of Plankton in a Lentic pond at Calcutta. In: Limnological Research in India (Ed. S. R. Mishra). Daya Publishing House, Delhi pp. 108-132.

Sawant R. S., P.D. Desai and J.S. Desai (2011). Physico-chemical properties of the Uttur Tank in Ajara tahsil of Maharasthra. *J. Ecobiol.* **29(3)**. : 205-211.

Saxena M.M.(1998). Environmental analysis. Water, soil and air. *Agro Botanica*, Bikaner, India.

Schlinder D.W. (1971). Carbon, Nitrogen, Phosphate and Eutrophication of freshwater Lakes. *J. Phycol.* **17**: 321-329.

Seenayya G. and Zafar A.R.(1979). An ecological study of Mir Alam lake, Hyderabad, India. *Ind.J.Bot.* **2(2)**. : 214-220.

Senthikumar R. and K, Sivakumar (2008). Studies on phytoplankton diversity in response to abiotic factors in Veeranam Lake in the Cuddalore District of Tamil Nadu. *J. Environ. Biol.* **29(5):** 747-752.

Sharma Dushyant (2006). Seasonal variations in different physico-chemical characteristics of Rampur Reservoir of Guna District (MP). In : Ecology of Lakes and Reservoirs (Ed. By V B Sakhare), Daya Publishing House,New Delhi. pp. 125-131.

Sharma Dushyant Kumar (2006). Seasonal variations in certain physico-chemical characteristics in Makroda Reservoir of Guna District. *Ecol. Envion. and Cons.* **7(2):** 201-204.

Sharma Dushyant Kumar (2007). Studies on the Ecology and Fish fauna of Makroda reservoir of Guna District (Madhya Pradesh). In Advances in aquatic Ecology Vol.I., Ed. V.B. Sakhare, Daya Publishing House, New Delhi. pp.31-34.

Sharma G.P. and G.P. Bhatnagar (1977). Seasonal variations in plankton biomass or organic matter in Bhopal Lake. India. *J.Zool. Soc. Ind.* **29:** 31-44.

Sharma P.K. and A.D.Adoni (1982). Studies on seasonal variations in pH and DO in Sagar lake. *Acta Botanica. India* **10** : 324-326.

Sharma Rekha and A.P. Diwan (1993). Limnological studies of Yeshwant Sagar Reservoir Plankton population dynamics. Recent Advances in freshwater Biology (Ed. K.S. Rao). Vol. I pp. 199-211.

Sharma Riddhi, Madhusudan Sharma, Vipur Sharma and Heena Malara (2008). Study of limnology and microbiology of Udaipur lakes. Proceeding of Taal 2007: The 12th world lake conference (Ed. M. Sengupta and R. Dalwani). pp. 1504-1508.

Sharma, Dushyant and Renu Jain (2000). Physico-chemical analysis of Gopalpura Tank of Guna District (MP). *Eco. Env. and Cons.* **6(4):** 441-445.

Shastri Yogesh (2005). Physico-chemical characteristics of a small percolation Tank. Advances in Limnology Ed. S.R. Mishra, Daya Publishing House, New Delhi. pp. 253-258.

Shastri Yogesh and D.C. Pendse (2001). Hydrobiological study of Dahikhuta reservoir. *J.Environ.Biol.***27(1):** 67-70.

Shastri Yogesh and P.K. Bhogaonkar (2006). Water quality of Talwade reservoir of Nashik District, Maharashtra. Ecology of Lakes and Reservoir Ed. V.B. Sakhare, Daya Publishing House, Delhi. pp. 138-142.

Shayestechfar A., M. Soleimani, S.N. Mousavi and F. Shirazi (2008). Ecological Study of rotifers from Kor River, Fars, Iran. *J. Environ. Bio.* **29(5)**: 715-720.

Shiddamallayya N. and M. Pratima (2007). Role of Limno-chemical factors in phytoplankton productivity. Advances in Aquatic Ecology Vol. 1 (Ed. V.B. Sakhare). Daya Publishing House, New Delhi. pp. 64-74.

Shinde, S. M and B. G, Kolhe (2014). Comparative Evaluation of Some Physico-chemical and Microbiological parameters in Godavari River Basin At Tapovan Areaof Nashik. *IJSR* **3(6):** 485-487.

Shukla S.C, B.D. Tripathi, Rajanikant, V. DeepaKumari and V.S.Pandey (1989). Physico-chemical and biological characteristics of river Ganga from Mirzapur to Ballia. *Indian J. Environ. Hlth* **31(3):** 218-227.

Singh Ajit and J.S. Laura (2012) An assessment of physico-chemical properties and phytoplankton density of Tilyar lake, Rohtak (Haryana) *Int. Jr.of Current Research* **4(05):** 047-051.

Singh A.K. and D.K. Singh (1999). A comparative study on the phytoplanktonic primary production of river Ganga and pond of Patna (Bihar), India. *J. Environ. Biol.* **20(3):** 263-270.

Singh B. and P.K. Saha (1987). Primary productivity in a composite fish culture pond at Kulia Fish Farm, Kalyani, W.B. *Proc. Nat. Acad. Sci.,* India. **57**: 11-25.

Singh G.S. and A.S. Singh (1994). Variation and correlation of Dissolved oxygen and effluent quantity and stage of river Ganga at Varanasi (India). *Indian J Environ. Hlth* **36(2)**: 79-83.

Singh H.P. (1998). Studies on primary production in Gobindsagar reservoir, Himachal Pradesh. *J. Environ. Biol.*, **19(2):** 167-170.

Singh M., B.K. Sinha and G. Sharma (1996). Dirunal migration of zooplankton in a fish pond of Patna, Bihar. *Geobios* **22**: 130-134.

Singh S.B. and R. Sahai (1978). Fluctuations in zooplanktons in relation to physico-chemical factors of a Pond. *Geobios* **5**: 228-230.

Singh, Ravindra (1990). Correlation between certain physico-chemical parameters and primary productivity in Jamalpur, Munger. *Geobios* **17**: 229-234.

Singhai S., G.M.A. Ramani and U.S. Gupta (1990). Seasonal variations and relationship of different physico-chemical characteristics in newly made Tawa reservoir. *Limnologica* (Berlin). **21(1):** 263-301.

Singhai S.G. (1986). Hydrological and ecological studies of newly made Tawa Reservoir at Raipur. Ph.D. Thesis, H.S. Gaur University, Sagar.

Sinha Sankar Narayan and Mranal Biswas (2011). Analysis of physico-chemical characteristics to study the water quality of lake in Kalyani, West Bengal. *Asian J. Exp. Biol. Sci.* **2(1):** 18-22.

Sivakumar K. and R. Karuppasamy (2008). Factors affecting productivity of phytoplankton in a Reservoir of Tamil Nadu, India. *Am-Euras J. Bot.* **1(3):** 99-103.

Smith G.M. (1950). The freshwater algae of united states, Mc Graw Hill Book Co. N.Y.

Solaraj G., S. Dhakumar, K.R. Murthy and R. Mohanraj (2010). Water quality in selected regions of Cauvery Delta River Basin, Southern India, with emphasis on monsoonal variation. *Environ. Monit. Aessess.* **166(1-4):** 435-444.

Somal S.K., B. Pradhan and T.N. Tiwari (2005). Water quality status of Hirakund Reservoir and its suitability for irrigation. Advances in Limnology Ed. S.R. Mishra, Daya Publishing House, New Delhi pp. 91-97.

Soruba R. and B. Sangeetha (2011). Phytoplankton diversity of the temple pond Minjur, Tiruvalluvar District. *J.Ecobiol* **29(4).** : 305-309.

Sreelatha K. Sree and S. Rajalakshmi (2008). Zooplankton Diversity of Goutami Godavai Estuary, Yanam, Pondicherry (UT). Advances in

Aquatic Ecology vol. II Ed. V.B. Sakhare, Daya Publishing House, New Delhi. pp. 38-42.

Sreenivasan A. (1964). Limnological features and primary production in a polluted moat at Vellore Madras state. *Environ. Biol.* **19(2):** 167-170.

Sreenivasan A. (1965). Limnology and productivity of tropical impoundment III. Limnology and production of Amaravati Reservoir Madras State, S. India. *Hydrobiologia* **26:** 501-516.

Sreenivasan A. (1968). Limnology of tropical impoundment of two hard water reservoirs in Madras state. *Arch.Hydrobiol.* **65:** 205-222.

Sreenivasan A. (1976). Limnological studies of primary production in three temple pond ecosystems. *Hydrobiologia* **48**: 117-123.

Srinivasaro S., A.M. Khan, Y.V.S.S. Lova Rami and M.V. Raghuram (2007). Variation of physical characteristics of Godavari River water at Nanded (Maharashtra). and Rajahumunotry (AP). *J. Aqua. Bio.* **22(2)**: 91-95.

Srivastava, Deepak Kumar and Naveen Krishna Srivastava(2013). Water Quality of Gomati Riverwater in and around Sultanpur of Uttar Pradesh, State, India. *Indian J.L.Sci.* **2(2)**. : 71-75.

Strom K.M. (1930). Limnological observations in Norwegion Lake. *Arch. Hydrobiol.* **21**: 97-127.

Subbamma D.V. and Ramma Sarma D.V. (1992). Studies on the water quality characteristics of a temple pond near Machillipatnam. Andhra Pradesh. *J. Aqua. Biol.* **7(1 and 2):** 22-27.

Subramanian P., R. Prabhakaran, K. Babu, S. Chidambaram, R.S. Kumar, B. Selvaraj M.,Senthilkumar, K. Srinivasamoorthy (2012)Groundwater quality investigation in Andimadam area, Perambalur district, Tamil Nadu, India. *Int. Jr.of Current Research* **4(05):** 168-172.

Sukand B.N. and H.S.Patil (2004). Water quality assessment of fort lake of Belgaum (Karnataka). with special reference to zooplankton. *J.Environ. Biol.* **25**: 99-102.

Sukumaran P.K. (2002). Primary production dynamics of a Perennial tank in Bangalore, Karnataka. *Geobios.* **29(1):** 41-46.

Sukumaran P.K. (2003). Diel variations in the biotic and abiotic parameters of a freshwater tank in Bangalore, Karnataka. *Geobios.* **30(4):** 277-283.

Swingle H.S. (1967). Standardization of chemical analysis of water and ponds muds. FAO Fish Rep. **44**: 337-342.

Synudeen Sahib S. (2011). Physico-chemical parameters and phytoplankton in the Parappar Reservoir, Kerala. *J. Eco. Biol.* **28(2)**: 187-190.

Takamura K., Y. Sugaya, N. Takamura, T. Hanazato, M. Yasuno and T. Iwakuma (1989). Primary production of phytoplankton and standing crops of zooplankton and zoobenthos in hypertrophic lake, Teganuma. *Hydrobiologia*. **173:** 173-184.

Thakur F.J., D.K. Bhol, Y.D. Parmar, A.B. Parmanad, M.B. Chauhan (2011). Seasonal variations in different physico-chemical parameters of Pariyej Lake, Kheda District, Gujrat. *Asian J. Environ. Sci.* **6(1):** 37-41.

Tiwari A. and S.V. Chauhan (2006). Seasonal phytoplanktonic diversity of Kitham Lake, Agra. *J. Environ. Biol.* **27(1):** 35-38.

Todd, D. K. (1995). Ground water Hydrology. John Wiley and Sons, New York.

Tripathi A.K., and S.N. Pandey (1990). Water Pollution. Ashish Publishing House, New Delhi.

Trivedy R.K. and P.K. Goel (1986). Chemical and Biological Methods for water pollution studies. Environmental Publications, Karad (MS).

Umamaheswri S. and N. Anbu Saravanan (2009). Water quality of Cauvery River Basin in Trichirappalli, India. *Int. J. of Lakes and Rivers.* **2(1):** 1-20.

Unni K.S. (1971). An ecological study of macrophyte vegetation of Doodhari Lake in Raipur. Part II. Physiological factors. *Hydrobiol.* **38(3):** 479-488.

Unni K.S. (1984). Limnology of sewage polluted tank in Central India. *Int. Rev. Hydrobiol.* **69(4)**. : 553-565.

Unni K.S. (1971). Comparative limnology of several reservoirs in Central India. *Int.Rev. Hydrobiol.* **70(6)**. : 845-856.

Unni K.S., A. Chauhan, M Varghese and L P Naik (1992). Preliminary hydrobiological study of river Narmada from Amarkantak to Jabalpur. In Aquatic Ecology.Ed. S R MIshra and D N Sakesana. Ashish Publishing House, New Delhi pp. 221-229.

Valecha V. and G.P. Bhatnagar (1989). Primary productivity of phytoplankton in a eutrophic lower lake Bhopal, India. *Envi and Ecol.* **7(1)**: 202-205.

Vardhan Harsh, A. K. Verma, Sartaj Ahmad Allayie and Raheela Mushtaq (2015). Assessing Variations in Physico-chemical parameters of Tatapani Spring of District Rajouri – Jammu. *Indian J.Sci.Res.* **11(1).** : 133-138.

Varghese M., A. Chauhan and L.P. Naik (1992). Hydrobiological studies of a domestically polluted tropical pond. 1. Physico-chemical characteristics. *Poll. Res.* **11(2):** 95-100.

Vasanth Kumar B. Pradeep V. Khajure and S.V. Roopa (2011). Aquachemistry, zooplankton and bacterial diversity in three ponds of Karwar District, Karnataka. *Rec. Res. Sci. Tech.* **3**: 39-48.

Vasanthkumar B. and K. Vijay Kumar (2011). Diurnal variation of physico-chemical properties and primary productivity of phytoplankton in Bheema River. *Rec. Res. Sci. Tech.* **3(4).** : 39-42.

Vass K.K., H.S. Raina, O. P. Zutschi and M A Khan (1977). Hydrobiological studies on river Jhelum. *Geobios* **4**: 238-242.

Verma P.K. and J.S. Dutta Munshi (1987). Plankton community structure of Bandra reservoir, Bhagalpur. *Tropic Ecol.* **28** : 200-207.

Verma M.N. (1969). Hydrobiological studies of a tropical impoundment Tekanpur reservoirGwalior, India with special reference to breeding of Indian carp. *Hydrobiologia* **34**: 358-368.

Verma S.R and G.R. Shukla (1968). Hydrobiological studies of a temple tank Devikund in Deoband (U.P.), India. *Indian J. Environ.Hlth* **10**: 177-188.

Vidya Gurkar M.S. and M.K. Mahesh (2011). Phytoplankton diversity in Varuna Lake of Mysore. *J.Ecobiol* **29(4).** : 299-304.

Vijay Kumar (1994). Seasonal variation in the primary productivity of a tropical pond. *J. Ecobiol.* **6**: 207-211.

Vijay Kumar K. (1995). Limnology of freshwater Pond of Gulbarga. Recent Researches on Aquatic Environment (Ed. Ashutosh Gautam and N.K. Agarwal). Daya Publishing House, New Delhi. pp. 87-97.

Vijay Kumar K., H. Anjinappa, Virupanna and C. Padmavathi (2000). Seasonal variations in the primary productivity of Gobbur tank, Gulbarga. *Proc. Acad. Environ. Biol.* **9(2):** 119-123.

Welch P.S. (1952). Limnology. Mc, Graw Hill Co. Inc. New York.

Wetzel R.C. (1975). Limnology, Philadelphia. USA.

WHO (1984). International standards for drinking water. World Health Organization Technical Report.

Wisharad S.K. and S.N. Mehrotra (1988): Periodicity and abundance of plankton in Gulasia reservoir in relation to certain physico-chemical conditions. *J. Ind. And Fish Soc.,* India. **20**: 42-49.

www.whoindia.org/./Water_Quality_neeri_Annexure-_DRINKING _WATER_-_SPECIFICATION.pdf.

Yadav R.C. and V.C. Srivastava (2011). Physico-chemical properties of water of river Ganga at Ghazipur. *Indian J. Sci. Res.* **2(4):** 41-44.

Yalavarthi, Eswari (2002). Hydrobiological studies of the red hills reservoir, North Chennai, Tamil Nadu. *J.Aqua. Biol.* **17(1):** 13-16.

Yousoff F.M. and I. Patimah (1994). A comparative study of phytoplankton population in two Malaysian lakes. *Mitt. Internat. Verein Limnol.* **24**: 251-257.

Zafar A.R. (1964). On the ecology of algae in fish ponds of Hyderabad, India I Physicochemical complexes. *Hydrobiologia* **23**: 179-195.

Zafar A.R. (1966). Limnology of Mussian Sagar lake Hyderabad, India. *Phykos*. **5**: 115-126.

Zafar A.R. (1967). On the ecology of algae in fish ponds of Hyderabad, India III The periodicity. *Hydrobiologia* **30**: 96-112.

Index

www.ingramcontent.com/pod-product-compliance
Ingram Content Group UK Ltd.
Pitfield, Milton Keynes, MK11 3LW, UK
UKHW021011290726
14059UKWH00001BA/76

9 789390 384273